# Linearly Polarized IR Spectroscopy

Theory and Applications for Structural Analysis

# LINEARLY POLARIZED IR SPECTROSCOPY

## Theory and Applications for Structural Analysis

BOJIDARKA IVANOVA
TSONKO KOLEV

CRC Press
Taylor & Francis Group
Boca Raton  London  New York

CRC Press is an imprint of the
Taylor & Francis Group, an **informa** business

CRC Press
Taylor & Francis Group
6000 Broken Sound Parkway NW, Suite 300
Boca Raton, FL 33487-2742

First issued in paperback 2019

ISBN 13: 978-1-138-11285-8 (pbk)
ISBN 13: 978-1-4398-2559-4 (hbk)

### Library of Congress Cataloging-in-Publication Data

Koleva, Bojidarka.
  Linearly polarized IR spectroscopy : theory and applications for structural analysis / authors, Bojidarka Koleva, Tsonko Kolev.
       p. cm.
    "A CRC title."
    Includes bibliographical references and index.
    ISBN 978-1-4398-2559-4 (hardcover : alk. paper)
    1. Infrared spectroscopy. 2. Polarization spectroscopy. 3. Materials--Analysis. 4. Chemical structure. I. Kolev, Tsonko. II. Title.

QD96.I5.K65 2010
541'.22--dc22
                                                                    2009049465

Visit the Taylor & Francis Web site at
http://www.taylorandfrancis.com

and the CRC Press Web site at
http://www.crcpress.com

# Contents

# Preface

The potential of polarized spectroscopy as a tool for experimental study of oriented compounds has been demonstrated for almost 120 years. The development of this method began as early as 1888 by Ambronn [1], who published the results obtained from the first linear-dichroic study on membranes. Since then, prominent publications, review articles, and books devoted to this spectacular research area have appeared. These investigations include wide-scope studies on polarized ultraviolet and vibrational (infrared [IR] and Raman) spectroscopy of oriented polymers, embedded compounds in stretched polymers, oriented single crystals, dissolved molecules in nematic liquid crystal hosts, polycrystalline samples oriented by melting, and membrane techniques. The fundamental work on IR spectroscopy of polymers by Zbinden [2], Frushour [3], Jasse and Köning [4], Thulstrup and Eggers [5], Michl and Thulstrup [6,7], Shrader [8], and Siesler [9] has significantly contributed to the development of this research field. The studies of a number of other scientists should also be mentioned, including: Laurent [10] and Samori [11,12] on the linear dichroism of solute molecules within micelles and liquid crystalline catalysis; Vogt [13] on oriented thin films; and the series of fundamental works of Jordanov [14–24] on oriented molecules in liquid crystal solutions (homogenic systems). The investigations of the latter have been continued by Arnaudov, Andreev, Tsankov, Kolev, and Rogojerov [17,25–31]. The theoretical background of polarization IR spectroscopy has been summarized in many of these fundamental contributions. Therefore, the polarization IR spectroscopic method is undergoing continued development, particularly with respect to the search for both new orientation approaches and theoretical generalizations. These contributions result in growing possibilities for the application of an inherently "classical" method such as IR spectroscopy for the instrumental structural characterization of compounds.

Colloid suspensions (heterogeneous systems) in nematic liquid crystal (CS-NLC) hosts were first employed in 2004 by Ivanova and her coworkers [32] as a tool for the partial orientation of solids to be examined by linear-dichroic infrared spectroscopy. The authors of this book have found that it is possible to perform spectroscopic and structural elucidation of the embedded compounds, independent of their melting point, crystalline or amorphous states, and the quality of single crystals or polycrystalline samples. The method can be used for structural elucidation of inorganic and organic compounds, including organic salts, metal complexes, and glasses. The vast research of Ivanova and her coworkers has also covered a number of important topics, such as the theoretical background of the orientation method, validation for accuracy, influence of the host on peak positions, integral absorbance of guest molecule bands, optimization of experimental conditions, morphology of suspended particles, the influence of the physical and chemical properties of the liquid crystal medium on the degree of orientation of suspended particles, studies on the velocity of shearing the suspension, effects of the extent of roughening KBr-plates on the degree of orientation, the influence of the space group on the orientation parameters, effects of nature and balance of the forces acting on suspended particles and their degree of orientation, and the building of corresponding mathematical models [33–35].

Today, the application of the CS-NLC method is associated with more than 350 compounds as analytical objects, including a small number of inorganic compounds; a large number of organic compounds such as all essential amino acids, their salts, and amides; a series of aliphatic and aromatic small homo- and heteropeptides, their salts, and metal complexes; heterocyclic compounds, such as substituted pyridines, quinolines, benzimidazoles, benzotriazoles, and their salts and metal complexes; aromatic amines; pharmaceutical products; alkaloids; and organic dyes. The method has also been successfully employed for obtaining local information on the structures of tellurite, vanadate, and borate glasses.

This book is the first to be devoted to both the theory and practical application of partially oriented colloidal systems in nematic liquid crystals for IR spectroscopic and structural elucidation of an embedded chemical in the solid state. The main goal of this work is to indicate the possibilities of application of this new orientation technique for structural characterization of various compounds. More specifically, this book is intended to show the ways of obtaining important information about the structure and spectral properties of oriented compounds. The examples included serve to demonstrate both the advantages of this technique and its limitations associated with some specific systems. Although, unlike the single-crystal X-ray diffraction, IR spectroscopy cannot be regarded as an "absolute" instrumental method for structural analysis, particular emphasis is placed on the application of the method for systems providing reliable and complete information compared to those requiring caution in structural evaluation without the support of additional instrumental methods.

This book is divided into two parts, including the theoretical basis of linear-dichroic infrared (IR-LD) spectroscopy (Chapter 1), the orientation method as colloid suspension in nematic hosts (Chapter 2), and practical aspects (Chapters 3 through 6) associated with the use of the method for the characterization of inorganic chemicals and glasses as well as various classes of organic compounds. In all of the examples, the scope and limitation of the method are discussed and summarized. Therefore, this text can serve as a useful source of information not only for specialists in IR spectroscopy but also for other scientists as well as Ph.D. students working in the field of structural analysis. It can also be successfully used by B.Sc. and M.Sc. students who attend courses for advanced physical methods of analysis.

# Acknowledgments

We would like to express our sincere gratitude to the Alexander von Humboldt Foundation (Germany) for its continuous support of our research. We are thankful for the research fellowships granted, which provided us with the unique opportunity to work with prominent scientists at recognized educational and research institutions including the University of Dortmund—Department of Organic and Structural Chemistry, and Professor Paul Bleckmann (fellowship granted to Tsonko Kolev, 1988–1989 and 1991–1992); and Ruhr University Bochum—Department of Analytical Chemistry, and Profesor William S. Sheldrick (fellowship granted to Bojidarka Koleva, 2003, 2007–2009).

We also gratefully acknowledge the kind assistance of Dr. Michael Spiteller, Managing Director of the Institute of Environmental Research, Technical University Dortmund (Germany), for the successful long-time collaboration, including the valuable discussions and excellent conditions for research and a creative atmosphere within the framework of several research projects supported by the Bundesministerium für Bildung und Forschung (BMBF, Germany), Deutscher Akademischer Austausch Dienst (DAAD, Germany), and Deutsche Forschungsgemeinschaft (DFG, Germany).

We greatly appreciate the kind assistance of Dr. William S. Sheldrick, head of the Department of Analytical Chemistry at Ruhr University Bochum (Germany), for our fruitful collaboration and creative discussions.

# Authors

**Tsonko Kolev, D.Sc.,** was born in 1948 in Gradnitsa, Bulgaria. He obtained his M.Sc. degree in chemistry in 1973 at Sofia University (St. Kl. Okhridski), Bulgaria, and his Ph.D. in organic chemistry in 1982 from the Institute of Organic Chemistry, Bulgarian Academy of Sciences. Dr. Kolev became associate professor in 1993 and earned his D.Sc. degree in chemistry in 2000. He became a full professor in 2004. His current position is head of the Department of Organic Chemistry at Plovdiv University (P. Hilendarski), Bulgaria. Since 2003, Dr. Kolev has been a permanent member of the Chemistry Faculty Council at Plovdiv University. He has been head of the group in Organic Structural Analysis at the Institute of Organic Chemistry, Bulgarian Academy of Sciences since 1994. Kolev has been a member of the Bulgarian Chemical Society since 1990 and a member of the American Chemical Society since 2005. He has also been a member of the nonprofit legal organization the Bulgarian Society for Organic and Metaloorganic Chemistry since 1987, and a member of the Alexander von Humboldt Union in Bulgaria. Dr. Kolev was also an Alexander von Humboldt Fellow from 1987 to 1988 and from 1990 to 1992 at the Department of Structural Chemistry, Dortmund University. He has been an invited lecturer at a number of different academic, research, and industrial organizations in Germany (Hamburg, 1998; Dortmund, 1994, 1996, 1997; Mainz, 1998, 1999; Münhen, 1998; Kaiserslautern, 2000, 2001, 2002), Italy (FIAT Research Centre, Tolmezo, 2001), Austria (Wien, 1990), Hungary (Research Institute, Budapest, 1996), Macedonia (Scopje, 2006), and Serbia (Belgrad, 2004, 2005; Novi Sad, 2005). Dr. Kolev has participated in more than 50 scientific conferences with 20 invited and 13 key lectures, and more than 30 oral presentations and posters. He has supervised nine Ph.D. students. Dr. Kolev is also a member of the Organization Committee of the International Summer School on X-ray Analysis, IR, and Raman Spectroscopy in Sofia (2004–2008) at the Centre of Applied Spectroscopy, sponsored by DAAD. He has authored more than 210 scientific publications, 1 review article, 2 book chapters, and 1 monograph. Dr. Kolev has two daughters, Martina and Tsonka.

**Bojidarka Ivanova, Ph.D.,** was born in 1974 in Sofia, Bulgaria. She received her M.Sc. degree in physical and theoretical chemistry in 1997 from Sofia University (St. Kl. Okhridski), Bulgaria. She earned her Ph.D. in analytical chemistry from Sofia University in 2001. Since 2003, Dr. Ivanova has been an associate professor in chemistry, holding the corresponding position at the Department of Analytical Chemistry at Sofia University. She is also the head of the Group of Molecular Spectroscopy. Dr. Ivanova has been a member of the Faculty of Chemistry Council at Sofia University from 1992 to 1998 and from 2001 to the present. She has also been a member of the International Union of Pure and Applied Chemistry (IUPAC) since 2004, the Bulgarian Union of Organic and Metaloorganic Chemistry (Bulgarian Academy of Sciences) and the Bulgarian Society of Chemistry Education, History and Philosophy (CE&HPC) since 2005, the Alexander von Humboldt Union in Bulgaria since 2003, and the American Chemical Society since 2007. Dr. Ivanova is a member of the Organization Committee of the International Summer School on X-ray Analysis, IR, and Raman Spectroscopy in Sofia (2007–2008) at the Centre of Applied Spectroscopy, sponsored by DAAD. She has supervised six Ph.D. students. Dr. Ivanova is the author or coauthor of more than 160 scientific publications, 2 review articles, 1 book chapter, and 1 book in the field of structural chemistry and spectroscopy. She received the Best Young Scientist for 2003 award granted by the Council of the University Rectors in Bulgaria. In 2009, Dr. Ivanova received the Pitagor Award by the Bulgarian Government and Ministry of Education as the Best Young Scientist in the Whole Area of Knowledge. She is an Alexander von Humbldt Fellow (2003–present). Dr. Ivanova has one daughter, Tsonka.

# Introduction

Colloidal suspensions in nematic liquid crystals (NLCs) have been studied extensively because of the wide interest for their application in telecommunications, the high-tech industry, and medicine. Fundamental research has been conducted by Soville [36], Gast [37], Poulin [38,39], Borstnik [40], Lev [41], and Feng [42]. NLCs have been described as new, attractive soft matter systems that arise through the combination of colloidal suspensions and liquid crystals (LCs) [38,39,41,43–48]. However, in linear-dichroic infrared (IR-LD) spectroscopy for infrared (IR) band assignment and structural elucidation of suspended particles, colloid suspensions were first applied and developed in 2004, using the model organic complex system of Ivanova and coworkers [32]. Until then, only three orientation techniques of solids had been described: (1) polymer orientation by stretching; (2) single-crystal growing followed by direct IR-LD measurements; and (3) polycrystalline samples oriented after melting the solid compounds between previously roughened potassium bromide (KBr) plates [2–9]. However, difficult technical problems are involved with these techniques, for example, growing a "good quality" single crystalline sample at the temperature of 170°C involved in the melting technique, which is a critical value for KBr plates. The orientation of solids as a colloidal suspension in an NLC (CS-NLC) avoids many of these problems. It is readily carried out and the samples are prepared as suspensions at room temperature. The technique is fast, easy to employ, and does not depend on the melting point of the compound under investigation, the quality of the single crystalline, or the polycrystalline sample. Its advantage lies not only in the possibility of experimental IR-band assignment of suspended compounds but also on the possibility of obtaining structural and local structural information in solids, where the application of single-crystal X-ray diffraction is impossible as, for example, is the case with amorphous compounds and glasses. Of course, we must note that the idea of obtaining the structural information by means of IR-LD spectroscopy has been previously presented, where, for instance, the determination of $\alpha$- and $\beta$-helices has been demonstrated on oriented polymers [3,4]. However, it is clear that by only using the oriented polymers we are limited in the investigation of compounds that do not possess these properties. The CS-NLC method gives structural and local structural information for polycrystalline and amorphous samples, glasses, organic compounds with high melting points like salts and metal complexes, and so forth. At this point, we have used the method to characterize about 350 organic compounds, their salts, some metal complexes, and glasses. The validation of the orientation tool for accuracy, precision, and the influence of the host on peak positions and integral absorbances of guest molecule bands has been reported as has the optimization of some experimental conditions including the number of scans, the roughening of the KBr pellets, the quantity of compound studied to be included in the liquid crystal medium, and the ratios of Lorentzian to Gaussian peak functions in the curve fitting procedure on the spectroscopic signal at five different frequencies. An experimental design for quantitative evaluation of the impact of four input factors has likewise been reported [33–35]. Also discussed are the fundamental questions concerning this orientation tool, that is, the morphology of the suspended particles, particle size, and the influence of the physical chemical properties of liquid crystal medium on the degree of orientation of suspended particles. The velocity of the shearing of the suspension, the degree of roughening of the KBr plates and its effect on the degree of orientation, have also been investigated. Other factors of importance include the influence of the space group on the orientation parameter of the compound studied, the nature and balance of the forces acting on the suspended particles and their degree of orientation, as well as the mathematical model, and so forth [35].

# 1 Linear-Dichroic Infrared (IR-LD) Spectroscopy

## *Background*

In the "classical" infrared (IR) spectroscopy approximation using the Wilson [49], Schrader [8], and Hollaš [50] books, the selective absorption of light has been explained by the redistribution of electrical charges for a harmonic vibration that corresponds to the difference in the energy levels between two stationary vibrational states of an irradiated molecule. The oscillating dipole generated interacts with the IR irradiation of the same frequency (or wavelength), which results in the emergence of the corresponding absorption band in the IR spectrum.

One prerequisite for the mathematical formalism describing these phenomena is that the amplitudes of molecular vibrations are extraordinarily small, compared to the length of the chemical bonds involved and the equilibrium values of valence angles. Therefore, the dipole moment of any molecule containing $N$ atoms can be expressed with the so-called harmonic approximation, by the first two members of the following expression:

$$\mu = \mu_0 + \sum_{i=1}^{3N-6} \left( \frac{\partial \mu}{\partial r} \right)_{r_0} Q_{i \cdots}, \tag{1.1}$$

where $\mu_0$ is the dipole moment of the equilibrium molecular configuration, and $Q_i$ is the normal coordinate. Describing the mode of the $i$-vibration of amplitude $a_i$ with maximum deviation from the equilibrium state:

$$Q_i = a_i \cos 2\pi \nu_i t \tag{1.2}$$

Combining Equations 1.1 and 1.2 produces the following equation:

$$\mu = \mu_0 + \sum_{i}^{3N-6} \left( \frac{\partial \mu}{\partial r} \right)_{r_0} a_i \cos 2\pi \nu_i t \tag{1.3}$$

According to the latter, active vibrations ($IR_a$) in the IR spectrum are those for which $(\partial \mu / \partial r) \neq 0$, that is, an addition ($\mu_0 \neq 0$) or generation ($\mu_0 = 0$) of an oscillating dipole moment with frequency $\nu_i$ occur. In the opposite case, $(\partial \mu / \partial r) = 0$, these vibrations are considered as inactive ($IR_{ia}$) (e.g., in the case of homonuclear two-atom molecules). Thus defined, the basic selection rule does not consider the stereochemistry aspects, determining the probabilities of IR transitions. The intensity of a band, expressed by the absorption $A_i$, is proportional to the square of the amplitude accompanying the change of the dipole moment:

$$A_i \sim \left[ \left( \frac{\partial \mu}{\partial r} \right)_{r_0} \right]^2 = \vec{M}_i^{\,2} \qquad (1.4)$$

The vector $\vec{M}_i$ is the *moment of transition*, which shows that the change of the dipole moment for the $i$-vibration is determined not only by its value but also by its direction. Therefore, $A_i$ depends on the orientation of $\vec{M}_i$ with respect to the oscillating vector $\vec{E}_i$ with frequency $v_i$, which characterizes the electrical component of the monochromatic IR light interacting with the molecule. The latter vector is located on a plane perpendicular to the irradiation beam trajectory. In other words, the intensity of $A_i$ is proportional to the square of the scalar product, as expressed below:

$$A_i \sim \left( \vec{M}_i \cdot \vec{E}_i \right)^2 = \vec{M}_i^{\,2} \cdot E_i^2 \cos^2 \theta \qquad (1.5)$$

Here $\theta$ is the angle, determined by the projection of $\vec{M}_i$ on the plane defined by $\vec{E}_i$. The maximum value of the amplitude $A_i$ is reached with mutually parallel orientations of both vectors ($\theta = 0°$, $\cos^2 \theta = 1$), whereas with perpendicular orientation of the latter ($\theta = 90°$, $\cos^2 \theta = 0$) the corresponding value is zero.

The emission of the irradiation generated by the IR source is not only polychromatic but also *nonpolarized* by its nature. This means that, represented as a flat wave with length $\lambda_i$, the $i$-th monochromatic beam is spread throughout all possible planes crossing its trajectory. Thus, it forms a cylindrical optical channel of a diameter equaling the amplitude of the light wave. For standard sample preparation techniques of IR analysis, such as capillary layer, solution, and suspension in Nujol, molecules are randomly distributed and for most of them $\vec{M}_i$ is effectively oriented with respect to some of the planes of the $\lambda_i$ ($v_i$)-optical channel. For each IR-active vibration, the corresponding band will be registered in the IR spectrum.

The *linearly polarized IR irradiation*, which is spread throughout one plane, can only be obtained by appropriate devices (polarizers) assembled next to the source of light in the spectrometer (Figure 1.1). If the sample studied is also fixed toward a certain direction, the intensity of the IR spectral bands will depend on the angle of rotation of the polarizer toward the director of orientation **n**. The corresponding IR spectra are therefore defined as linear-dichroic infrared (IR-LD) ones. As a rule, they are registered as spectral curves (or as the corresponding subtractions derived from them) obtained at two different orientations of the polarized irradiation, most frequently parallel ($IR_p$) and perpendicular ($IR_s$) with respect to the macroscopic director **n** (Figure 1.1). The symbols $p$ and $s$ (in German, *parallel* and *senkrecht*; Born [51]) are equivalent to the designations ‖ and ⊥ and, therefore, have been adopted in the terminology of the present work.

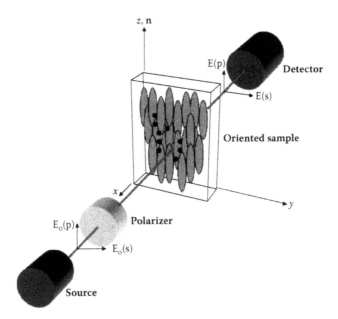

**FIGURE 1.1** Optical scheme for a measurement of IR-LD spectra: source of irradiation, polarizer, oriented sample, and detector. **(See color insert.)**

## 1.1 THEORETICAL PREREQUISITES

In the classical approximation mentioned in the previous section, the moment of transition, $M_i$, is formed as a vector sum from the amplitudes of deviation for the atoms involved in $i$-vibration [6,7]. In this way, it is determined to be, with respect to a coordinate system, associated with the molecule. For the oriented molecules of a given compound studied, however, the direction of $M_i$ is defined toward the director **n**, originating from the macroscopic orientation of the molecular assembly.

Figure 1.2 outlines the idealized case, for which the molecular and the macroscopic Descartes coordinate systems coincide. The plane of the IR-LD irradiation and the director **n** are parallel toward the z axis. The oriented molecules are of a single-axis mode, which suggests that their equilibrium geometry is associated with axial symmetry. Such molecules are $CO_2$, $H\text{-}C\equiv N$, and $CH_3\text{-}C\equiv C\text{-}CH_3$, which possess nondistinguishable configurations in their rotation around the corresponding axis. Many of the so-called real, strongly elongated (anisometric) molecules, for example, the nematic liquid crystals of a bicyclohexyl structural type (see later), can be regarded as moieties, possessing only a single axis. According to the model assumed earlier, the configurational distribution of such anisometric molecules with stereo structure is "averaged" with respect to a plane, which is perpendicular toward the director **n**, formed by the orientation of their long axis.

Based on these prerequisites, expressed in Descartes coordinates, the moment of transition $\vec{M}_i$ could be characterized by its components $\vec{M}_x$, $\vec{M}_y$, and $\vec{M}_z$ (Figure 1.2).

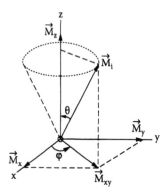

**FIGURE 1.2**  Definition of the moment of transition $\vec{M}_i$.

For the transition in spherical coordinates, we have:

$$\vec{M}_z = \vec{M}_i \cdot \cos\theta \tag{1.6}$$

$$\vec{M}_x = \vec{M}_i \cdot \sin\theta \cdot \cos\varphi \tag{1.7}$$

$$\vec{M}_y = \vec{M}_i \cdot \sin\theta \cdot \sin\varphi \tag{1.8}$$

and, respectively,

$$M_z^2 = M_i^2 \cdot \cos^2\theta \tag{1.9}$$

$$M_x^2 = M_x^2 \cdot \sin^2\theta \cdot \cos^2\varphi \tag{1.10}$$

$$M_y^2 = M_i^2 \cdot \sin^2\theta \cdot \sin^2\varphi \tag{1.11}$$

Since the molecules forming macro orientation are randomly rotated with respect to their long axis, coinciding with the director **n**, from Equation 1.9 it follows that the sum of the moments of transition $\vec{M}_i$ forms a conical surface with parameters ($\vec{M}_i$, $\theta$) of the same values of component $\vec{M}_z$. The base of this cone projects a circle on the four quadrants of the $xy$ plane with a center, located at the beginning of the laboratory coordinate system and with radius $\vec{M}_{xy}$ (Figure 1.2).

Consequently

$$(\vec{M}_x)^2 + (\vec{M}_y)^2 = (\vec{M}_{xy})^2 \tag{1.12}$$

and in spherical coordinates (see Equations 1.10 and 1.11) it can be expressed as follows:

$$(\vec{M}_{xy})^2 = \vec{M}_i^2 \cdot \sin^2\theta \cdot \cos^2\varphi + \vec{M}_i^2 \cdot \sin^2\theta \cdot \sin^2\varphi, \tag{1.13}$$

$$(\vec{M}_{xy})^2 = \vec{M}_i^2 \cdot \sin^2\theta (\cos^2\varphi + \sin^2\varphi) \tag{1.14}$$

Since $\cos^2\varphi + \sin^2\varphi = 1$, then

$$(\vec{M}_{xy})^2 = \vec{M}_i^2 \cdot \sin^2\theta \tag{1.15}$$

or, finally

$$\frac{(\vec{M}_z)^2}{(\vec{M}_{xy})^2} = \frac{\vec{M}_i^2 \cdot \cos^2\theta}{\vec{M}_i^2 \cdot \sin^2\theta} = \cot^2\theta \tag{1.16}$$

Bear in mind that the plane of the linearly polarized irradiation is oriented along the z axis and $(\vec{M}_z)^2$ and $(\vec{M}_{xy})^2$ are proportional to the absorptions, corresponding to the i-band in the parallel $(A_p^i)$ and the perpendicular $(A_s^i)$ spectrum.

The experimentally determined value

$$D_i = \frac{A_p^i}{A_s^i} \tag{1.17}$$

is called the dichroic ratio. From Equation 1.16, it follows that $D_i$ is only determined by the angle $\theta$, which means that all vibrations with the moment of transition forming a fixed angle with the orientation director **n**, have the same dichroic ratio. This conclusion is of principal importance for the application of IR-LD spectroscopy in structural analysis.

According to Equation 1.16, the mutual orientation of $\vec{M}_i$ and the director **n** concluding angle of 45° corresponds to the dichroic ratio $D_i = 1$, which means that the intensity of the band i should be the same in the parallel as well as in the perpendicular IR-LD spectrum. This, however, is not in fact the case, since $A_p$ and $A_s$ actually become equal at the *magic angle* of 54.7°.

This nonagreement can be explained by the model approximation assumed previously and the assumed perfect orientation of the molecules along the axis z ≡ **n**. Even for such idealized macroorientation, however, the thermal fluctuation of the molecules should generate an "averaged" axis along the z direction. This effect is evaluated by the introduction of the *parameter of order S* $(0 < S < 1)$. The remarkable work of Saupe and Maier must be mentioned in this respect [52–55]. Their theory (Maier–Saupe theory) of NLC is founded on a molecular field treatment of long-range contributions to the intermolecular potential and ignores the important short-range forces. Nonetheless, this theory is particularly successful in predicting the orientational properties of real nematics [44].

Figure 1.3 is associated with the moment of transition $\vec{M}_i$ oriented at an angle $\theta$ with the axis z′ in a coordinate system, along which the long molecular axis is oriented. If the director **n** coincides with the z axis of laboratory coordinate system

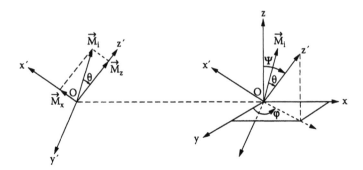

**FIGURE 1.3**   $\vec{M}_i$ involved in the conclusion of angle $\theta$ with the z' axis.

(x, y, z), the transformations indicated in Scheme 2 with the involvement of the *Euler* angles $\psi$ and $\varphi$ can be applied.

In this scheme, $\psi$ represents the angle of procession between axis OZ' and axis z (z-type axis) for the two coordinate systems ($0 \le \psi \le \pi$); $\Phi$ is the angle of free rotation, restricted by the axis y and the projection of z on the plane xy.

Regarding the molecular coordinate system, the following equations are applied:

$$\left(\vec{M}_{z'}\right)^2 = \vec{M}_i^2 \cdot \cos^2 \theta \tag{1.18}$$

$$\left(\vec{M}_{x'}\right)^2 = \vec{M}_i^2 \cdot \sin^2 \theta \tag{1.19}$$

and Equation 1.19 describes the projection of $\vec{M}_{x'y'}$ on the axis x' (Equation 1.7 and Figure 1.2).

The transformation $\vec{M}_{z'} \to \vec{M}_z (z \equiv \mathbf{n})$ can be obtained from the following expression:

$$\left(\vec{M}_z\right)^2 = \vec{M}_i^2 \left(\sin^2 \theta \cdot \sin^2 \psi \cdot \sin^2 \varphi + \cos^2 \theta \cdot \cos^2 \psi\right) \tag{1.20}$$

Bearing in mind the cylindrical symmetry of the oriented molecules and their uniform distribution with respect to the orienting z axis ($0 \le \varphi \le 2\pi$), the averaged value of $\sin^2\varphi$ can be expressed as follows:

$$\overline{\sin^2 \varphi} = \frac{1}{2} \tag{1.21}$$

On the other hand, the averaged values $\overline{\sin^2 \psi}$ and $\overline{\cos^2 \psi}$ are associated with the parameter of order $S$ by the following relationships:

$$\overline{\sin^2 \psi} = \frac{2}{3}(1-S); \quad \overline{\cos^2 \psi} = \frac{1}{3}(2S+1)$$ (1.22)

By using the expression $\cos^2\theta = 1 - \sin^2\theta$ from Equations 1.20 and 1.21, the next equation can be obtained:

$$\left(\vec{M}_z\right)^2 = \frac{1}{3}\vec{M}_i^2[2S(1-\frac{3}{2}\sin^2\theta)+1]$$ (1.23)

If, again, the uniform distribution of the oriented single-axis molecules with respect to the $z \equiv \mathbf{n}$ is considered, it can be assumed that $\vec{M}_x$ and $\vec{M}_y$ contribute equally to the perpendicular polarization. Since

$$\left(\vec{M}_x\right)^2 + \left(\vec{M}_y\right)^2 + \left(\vec{M}_z\right)^2 = \vec{M}_i^2$$ (1.24)

Therefore,

$$\left(\vec{M}_x\right)^2 = \frac{1}{2}\left[\vec{M}_i^2 - \left(\vec{M}_z\right)^2\right]$$ (1.25)

According to the model assumed, both the parallel ($IR_p$) and perpendicular ($IR_s$) linear dichroic spectra correspond to orientation of the polarized irradiation along the axis $z \equiv \mathbf{n}$ and the axis $x$, respectively. For the dichroic ratio of $D = 1$, we have:

$$A_p = A_s \sim \left(\vec{M}_z\right)^2 = \left(\vec{M}_x\right)^2$$ (1.26)

According to Equations 1.23 and 1.25, this means that the following expression can be applied:

$$2S\left(1-\frac{3}{2}\sin^2\theta\right) = 0$$ (1.27)

or

$$\frac{3}{2}\sin^2\theta = 1 \rightarrow \sin^2\theta = \frac{2}{3}$$ (1.28)

The corresponding angle $\theta$ is equal to the "magic angle" of 54.7° (Figure 1.4).

The physical meaning involved in the parameter of order $S$ appears to be more universal than that assumed merely by its participation in the relationships discussed earlier and the determination of the magic angle value. This meaning reflects the

**FIGURE 1.4**   Definition of angle θ.

fact that polarization measurements are only possible if a sample of a certain degree of orientation ($S \neq 0$) is available. A higher value of the parameter of order means a registration of higher-quality IR-LD spectra, which increases the reliability of the results obtained. For the NLCs, which are used as an orientation medium in the method presented in this book, and at room temperature, the value of $S$ is between 0.3 and 0.5. For compounds with lower temperature limits of the mesomorphic interval, the corresponding value of ~0.8 can be reached [56].

From this discussion concerning the theory of the linear IR dichroism, the following generalizations can be postulated: If, by employing appropriate methods, a single-axis macro configuration of the molecules of a compound studied is statistically fixed, the polarization of its IR spectral bands will depend on the orientation determined by the angle $\theta_i$ of the corresponding moments of transition $\vec{M}_i$ with respect to the macroscopic director **n** (for the basic vibrations $i \in \{1, 3N - 6\}$). The translation toward the beginning of the molecular coordinate system of the subset of $\vec{M}_j$ vectors ($j < i$) with $\theta_j$ = constant, forms a conical surface of the second order (i.e., double cone), since the orientation **n** determines *a direction as a whole, not simply a relative direction*. For the $\vec{M}_j$ moments, projected on the plane, irrespective of their values, the corresponding IR-LD bands of the same dichroic ratio can be obtained.

Moments of transition that are colinearly oriented with respect to the director **n**, to which bands of the same dichroic ratio and the same response toward the polarization measurement correspond, are generated by vibrations of the same type of symmetry. This is why the area of successful application of the IR-LD spectroscopy for structural characterization can be significantly expanded by preliminary analysis of the mode of molecular vibrations and their grouping, according to identical types of symmetry. From the practical point of view, the efficiency of the method depends also on the resulting orientation of a sample as well as the mode of processing the IR-LD spectral data obtained.

## 1.2   SYMMETRY ANALYSIS OF NORMAL VIBRATIONS AND DIPOLE MOMENTS OF TRANSITION GENERATED THEREFROM

Some principal terms can be introduced when discussing the symmetry analysis of the molecular (point) symmetry of a given chemical compound [57,58].

### 1.2.1   ELEMENTS AND OPERATIONS OF SYMMETRY

Elements and operations of symmetry are interrelated terms, that is, a certain operation of symmetry is applied to a specific element, which results in the adoption of a molecular configuration that cannot be distinguished from the initial one. Some principal terms in this respect are listed and explained in the following:

> If *rotation axis of symmetry* $(C_n)$ exists, the initial configuration is recovered after rotation (*operation of symmetry*) at the angle of $2\pi/n$ around a line crossing a molecule (*element of symmetry*). Linear molecules generate $C_\infty$-axis of symmetry automatically.
>
> The *plane of symmetry* $(\sigma)$ crosses a molecule. The reflection from the plane of symmetry (*operation of symmetry*) causes an exchange of the locations of the equivalent atoms. A planar molecule by itself is associated with the existence of plane of symmetry.
>
> *Center of symmetry* $(i)$—The initial configuration is recovered by reflection of a center, included in the molecular geometry.
>
> *Mirror-rotation axis of symmetry* $(S_n)$—The rotation around a molecular $C_n$-axis at the angle of $2\pi/n$ and the following reflection on a plane, which is perpendicular toward this axis (a *composite operation of symmetry*).
>
> *Classification of the basic (normal) vibrations, according to their mode of symmetry*—The operations of symmetry change specifically the mode of normal vibrations in a molecule. In this respect, a certain classification characterizes these vibrations according to their type of symmetry with respect to the particular operation and/or element of symmetry that can be presented.

Normal vibration, which does not change its mode as some operation of symmetry is applied, is defined as symmetrical (s). If the operation of symmetry reverses the sign of coordinates (−1) determining the mode of vibration, the latter is defined as nonsymmetrical (as). The third possibility is to generate a new mode of transition, and such vibrations are called degenerative (e) vibrations.

For the full characterization of the symmetry mode of molecular vibrations, additional, not entirely standardized symbols are introduced. Nondegenerate vibrations are designated by the capital letters *A* and *B*, and these vibrations are, respectively, symmetrical and nonsymmetrical toward the rotating $C_n$ axis. The letters *E* and *F* correspond to double and triple type, respectively, degenerated vibration with rotation around this axis. The subscripts 1 and 2 indicate, respectively, symmetry and anti-symmetry with respect to a plane, which includes a rotating axis of symmetry. The small letters *g* and *u* determine the type of symmetry with respect to a center of symmetry ($g \equiv s$ and $u \equiv as$). Upper lines are used to indicate symmetry (single line) and antisymmetry (double lines) with respect to a plane, which is perpendicular toward a given axis of symmetry.

Figure 1.5 and Table 1.1 are associated with the elements, operations, and classification on the basis of the symmetry mode for a nonlinear molecular fragment of the type $XY_2$. Such a geometry is characteristic for organic compounds, containing substituents or functional groups such as $>CH_2$, $-NH_2$, and $-COO^-$. Considering the rest of the molecule as an effective "heavy atom," it can be assumed that these functionalities give rise to $4 \times 3 = 12 - 6 = 6$ localized vibrations of various shape, corresponding to the characteristic vibrations in the IR spectrum (Figures 1.5 and 1.6).

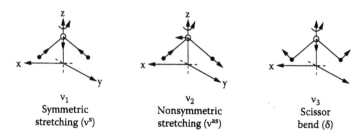

$v_1$
Symmetric
stretching ($v^s$)

$v_2$
Nonsymmetric
stretching ($v^{as}$)

$v_3$
Scissor
bend ($\delta$)

**FIGURE 1.5**   Normal vibrations for $xy_2$-type nonlinear molecules.

**TABLE 1.1**

**Type of Symmetry for the Basic Vibrations of $xy_2$ Fragments**

| Normal Vibration | Element of Symmetry | | | Type of Symmetry |
|---|---|---|---|---|
| | $C_2^z$ | $\sigma_{v(yz)}$ | $\sigma_{v(xz)}$ | |
| $V^s$ | + | + | + | $A_1$ |
| $V^{as}$ | − | − | + | $B_2$ |
| $\delta$ | + | + | + | $A_1$ |

If analysis of the three basic vibrations shown on Figure 1.5, which are typical for the possible modes of vibrations for nonlinear $xy_2$ molecules such as $H_2O$, is only conducted, the following elements, operations, and types of symmetry are involved:

| $v_1$ | $v_2$ | $v_3$ |
|---|---|---|
| symmetric | nonsymmetric | scissor |
| stretching ($v^s$) | stretching ($v^{as}$) | bend ($\delta$) |

The nonlinear $XY_2$-type molecules (more generally, the corresponding $XY_2$ structural fragments) possess three elements of symmetry:

1. $C_2$ rotating axis, oriented along the $z$ axis of the molecular coordinate system.
2. $\sigma_{xz}$ plane, formed by the equilibrium configuration of the $XY_2$–molecules.
3. $\sigma_{yz}$ plane, which is perpendicular to $\sigma_{xz}$, half-separating the molecule.

Both planes include the $C_2^z$ axis, and, more specifically, this axis is formed by their intersection.

The operations with respect to these three elements of symmetry rearrange the mode of the three vibrations. The $v_1$ stretching vibration (Figure 1.5), for which the two XY bonds are continuously extended or shortened with the same phase of vibration, does not change its mode with both the rotation at 180° around the $C_2^z$ axis and the reflection on the $\sigma_{yz}$ plane. In both cases, an exchange in the location of the equivalent Y atoms takes place, whereas the X atom remains on the z axis. Even if formally conducted, the third operation, which is associated with the reflection on $\sigma_{xz}$, leads to the same result. It coincides with the *operation of identity* ($I$), which does not cause any rearrangements. The generalization of these results indicates that the mode of $v_1$ can be characterized as belonging to the $A_1$-symmetry type.

For the $v_2$-stretching vibration, the vectors formed by the amplitudes of the Y atoms have opposite signs, that is, whereas one bond is extended, the other is shortened (Figure 1.5). The mode of $v_2$ is not changed only when its trivial reflection by the own $\sigma_{xz}$ plane takes place. The other two operations—the rotation around $C_2^z$ and the reflection by $\sigma_{yz}$—cause changes in the opposite direction (the corresponding sign becomes –1), for the coordinates of the two Y atoms as well as the X atom in the direction, which is perpendicular toward $C_2^z$. According to the nomenclature adopted earlier, this vibration can be characterized as $B_2$-type vibration, that is, it is antisymmetric with respect to a given axis and the corresponding plane of symmetry.

The planar $v_3$-type vibration of scissor-bending mode, designated by the symbol $\delta$, takes place in the plane $\sigma_{xz}$ itself (Figure 1.5). The response of its mode with respect to the three operations of symmetry is the same as that of the $v_1$-stretching vibration; therefore, it can also be characterized as a vibration of the $A_1$ symmetry type.

The aforementioned results are summarized in Table 1.1. The first and the last column indicate the designations of the corresponding vibrations and their types of symmetry. The symbols +1 and –1 are associated with symmetric and antisymmetric rearrangements, respectively, and show that the operation changes (or does not change) the sign of the vector deviations, determining a given mode of vibration; the amplitude of the latter, however, is not changed.

Table 1.1 is associated with a subgroup belonging to the $C_{2v}$-point group of symmetry. It can be expanded by the symmetry types of the remaining three normal vibrations for the generalized $XY_2$-structural fragment. The corresponding modes of vibration are shown in Figure 1.6.

| ρ rocking | ω wagging | τ twisting |
|-----------|-----------|------------|
| ($B_2$ class) | ($B_2$ class) | ($A_2$ class) |

Vibrations that are rearranged according to the same type of symmetry belong to the same *symmetry class*. In the specific case discussed, $v^s$ and $\delta$ vibrations are class $A_1$, and $v^{as}$ corresponds to class $B_2$. For the expanded version of the structural fragment $XY_2$, the rocking (ρ) as well as out-of-plane wagging (ω) vibrations are of the same class. The twisting (τ) vibrations belong to a separate $A_2$ class (Figure 1.6).

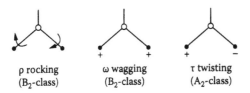

ρ rocking          ω wagging          τ twisting
($B_2$-class)       ($B_2$-class)       ($A_2$-class)

**FIGURE 1.6**   Additional bending vibrations for fragments of the $XY_2$ structural type.

## 1.2.2  SYMMETRY ANALYSIS OF THE DIPOLE MOMENTS OF TRANSITION

The operations of symmetry are also applicable to the moments of transition, generated by the normal vibrations of an $XY_2$ fragment. The corresponding symmetry analysis for $v^s$-stretching, $v^{as}$-stretching, and $\delta$-bending vibrations is shown (see Figure 1.7). The results obtained can be also generalized for molecules of more complex point-type symmetry. The directions of the three transition moments have been obtained by the conventional way of summing the vector deviations of atoms. The zero point corresponds to the center of masses (the beginning of the molecular coordinate system), which in this case coincides with the equilibrium coordinate of the heavy X atom and is located along the z axis. The resultant transition moments are shown on the right side of each of the drawings below (Figure 1.7).

As seen from Figure 1.7, the moments corresponding to the $v^s$-stretching and the $\delta$-bending vibrations are oriented along the z axis. This means that they are invariant (+1) with respect to the three symmetry operations and, therefore, belong to the symmetry class $A_1$. The moment of transition derived by $v^{as}$ is located on the $\sigma_{xz}$ plane and is directed perpendicularly with respect to the z axis. Such orientation corresponds to the symmetry class $B_2$, since its coordinate changes its sign (–1) on rotation at 180° around the z axis and its reflection by the $\sigma_{yz}$ plane.

This analysis illustrates a particular case of one fundamental IR spectroscopy rule: The dipole moment of transition is characterized by the same symmetry mode as the normal vibration, which generates it [57]. This means that colinear moments of transition and bands of equal dichroic ratio correspond to vibrations belonging to the same symmetry class. This conclusion is of principal importance for the application of IR-LD spectroscopy for the identification of the IR bands. The statement is supported by the example given in Figure 1.8.

The simplified case of the $XY_2$-type molecular fragment is illustrated by Figure 1.8; however, this molecular fragment belongs to an oriented molecule, and the z axis coincides with both the director of macro orientation **n** and the Y–X–Y angle bisector. Under these prerequisites, the $v^s$ and $\delta$ bands have the maximum intensity if the polarized irradiation is spread throughout the xz plane, since the moments of transition generating these bands coincide with the vector $\vec{E}_i$ of the light wave (Figure 1.8b). The $v^{as}$-type band, corresponding to the transition moment perpendicularly oriented toward this plane, will be not registered whatsoever. The opposite effect will be observed if the polarized irradiation is restricted within the xy plane (Figure 1.8c).

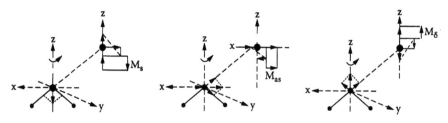

**FIGURE 1.7**  Direction of the dipole moment of transition for $v^s$-stretching, $v^{as}$-stretching, and $\delta$-bending vibrations.

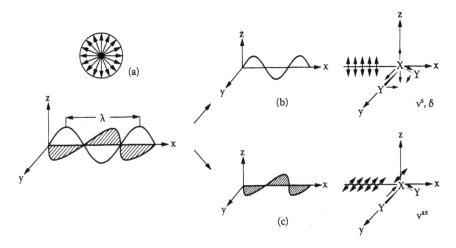

**FIGURE 1.8** Analysis of the polarization of the principal bands for $XY_2$-type molecular fragment: (a) nonpolarized IR irradiation; (b) parallel polarization; and (c) perpendicular polarization.

Then the bands, corresponding to both the stretching symmetric and bending $XY_2$-type vibrations will disappear, since they will be *mutually eliminated*.

It is indicated that the simultaneous elimination of bands belonging to a common symmetry class, that is, bands of the same dichroic ratio, can be achieved by *differential reduction*.

## 1.3  ORIENTATION OF THE SAMPLES

The theoretical model of the IR linear dichroism suggests a single-axis orientation of a molecular assembly with axial (cylindrical) symmetry. This means that the molecules are located with their longest axis in the direction, which induces the anisotropy, determined by the director **n**, and the side groups are randomly fixed or rotate along the length of this axis. Under these conditions, the dipole moments of transition caused by the normal vibrations form conical surfaces with director **n** (Figure 1.2) [2,3,6–8].

The macro structure of a single crystal with director **n**, determined by one of the crystallographic axes, represents the idealized example of an oriented system, for which the axial symmetry is replaced by the translation one. IR-LD spectral studies of single crystalline samples have been conducted first and were widely applied for structural characterization. Regardless of its vast use, the method is characterized by a number of complications. In many cases, the preparation of a sample of appropriate quality and dimensions is associated with some technical problems. The detailed analysis of the IR-LD spectral data is also complicated, since, as a rule, the elementary cell contains more than one molecule, and these molecules participate in the formation of regular chains of the crystalline lattice, which are noncolinearly located with respect to each other. The effect from this is the presence of bands of the same origin, strongly superimposed toward each other, which are mutually eliminated at different dichroic ratios. Additional splitting of the absorption maxima is also possible because of both the static and dynamical effects exerted by the crystalline field [59].

The preparation and the study of oriented polycrystalline layers are believed to circumvent the difficulties associated with the preparation of monocrystalline samples. The method is based on the crystallization of a film out of a melt, located between potassium bromide (KBr) plates, which are transparent with respect to the irradiation in the middle IR region. If certain conditions of temperature gradient are observed, such as slow cooling of the sample, relatively good *partial* orientation can be achieved [6,7]. This method, however, has its limitations. The interpretation of the IR-LD spectra is also complicated, as it is in the case of monocrystalline samples. Moreover, the preparation of oriented layers is only possible for samples with a melting point below 160°C to 170°C, since higher temperatures would damage the KBr plates. For this reason, most of the inorganic and coordination compounds as well as organic salts that decompose on melting cannot be subject to the orientation procedure.

Studies of polymer films, oriented by extension toward a certain direction, have also been widely employed, since essential information about the structure of the corresponding high-molecular compounds can be obtained by employing this procedure [2,9].

The aforementioned three methods ensure the appropriate orientation of macro systems; however, as already pointed out, the corresponding IR-LD spectra are complicated by superimposing with data concerning the super molecular structure. The problem associated with the interpretation of the IR-LD spectra and the structural characterization of *individual molecules* has led to the development of two orientation methods: (1) orientation within a polymer matrix, and (2) orientation in a liquid crystal solution. The first method is associated with the introduction of the "guest molecules" from the compound studied into a polymer film, prepared by the dissolution of a mixture of poly(ethylene) and the substance studied in toluene or chloroform. After evaporation of the solvent, a thin film is prepared that is consequently subjected to multiple (hundredfold or more) procedures of extension. The resulting orientation of the polymer chains induces orientation of the guest molecules as well. There are two principal reasons for the usage of poly(ethylene) as a matrix. The first one is its chemical inertness, and the second is the transparency of poly(ethylene) with respect to irradiation in the middle IR region, except for the basic absorption intervals for alkyl groups at 3000 to 2800 cm$^{-1}$, 1480 to 1440 cm$^{-1}$, and 750 to 720 cm$^{-1}$.

The technique for orientation in a polymer matrix introduced initially for studying electron transitions [5] has been widely employed for the characterization of the IR-LD spectra of organic compounds [60–64]. The disadvantages of the method are the relatively long time necessary for the preparation of samples and diffusion phenomena associated with the compound studied, which result in changes in its concentration in the matrix during the course of time.

## 1.3.1 Orientation in Liquid Crystal Solutions

Liquid crystals (LCs) possess properties of the matter between that of a crystalline and an isotropic liquid. Many of the mechanical properties of liquid, such as high fluidity, formation, and coalescence of droplets, are retained. At the same time they are similar to crystals in that they exhibit anisotropy in their optical, mechanical, electrical, and magnetic properties.

There are many different types of LC phases, which can be distinguished based on their different optical properties. LCs can be classified as thermotropic, lyotropic, and metallotropic. Thermotropic LCs exhibit a phase transition into the LC phase as temperature is changed. The rigid part of a liquid crystal molecule is called a mesogen. Lyotropic LCs exhibit phase transitions as a function of concentration of the mesogen in a solvent. Liquid crystal phases can also be based on low-melting *inorganic* phases like $ZnCl_2$ that have a structure formed of linked tetrahedra and easily form glasses. The addition of long chain soap-like molecules leads to a series of new phases that show a variety of liquid crystalline behavior both as a function of the inorganic–organic composition ratio and of temperature. This class of materials has been named metallotropic. On the other side the thermotropic liquid crystals can be divided into nematic phases, smectic phases, chiral phases, blue phases, and discotic phases. Nematic liquid crystals are the simplest mesophases, where the long molecular axes are preferably oriented in one direction, defined as director **n** (Figure 1.4). The molecular dipoles are compensated and in equilibrium the nematic mesophase are electrically neutral; that is, both directions of the director, **+n** and **−n**, are equivalent. The ordinary nematic structure shows an optically positive uniaxial behavior with the optical axis parallel to director **n**.

The word *nematic* comes from the Greek word νημα (*nema*), which means "thread." This term originates from the thread-like topological defects observed in nematics, which are formally called disclinations. Nematics also exhibit so-called hedgehog topological defects.

Nematics have fluidity similar to that of ordinary (isotropic) liquids but they can be easily aligned by an external magnetic or electric field. Aligned nematics have the optical properties of a uniaxial crystal and this makes them extremely useful in liquid crystal displays (LCDs).

*Nematic liquid crystals* as thermotropic systems are also employed as orienting medium in the IR-LD analysis [65,66]. Thermotropy means that the liquid-crystal phase (mesophase) is implemented within a certain interval between the melting temperature and the *point of clarification*. Within these temperature limits, the liquid crystal molecules are located toward a certain direction, inducing orientation without any positional order (Figure 1.9). In their mesomorphic interval, the nematic liquid crystals are characterized by their lower fluidity and transparency as well as anisotropy in their optical properties such as the double-beam refraction and dichroism.

Molecules of elongated (anisometric) shape, consisting of "rigid nucleus" (such as biphenyl, benzylideneaniline, azocyclohexyl, and bicyclohexyl) fragments, and flexible "tails" (such as alkyl, alkoxy, carboxyl, ester, and other functionalities) are known to form in the nematic phase. Both the polar and dispersion forces of attraction between the rigid fragments of these molecules are counterbalanced by the steric effect of the flexible tails, which cause the necessary monoaxial orientation; moreover, the flexible end groups provide an additional fluidity. The effect of thermal fluctuation, which is typical for these systems, is more pronounced at

**FIGURE 1.9**   Scheme of a nematic liquid crystal.

higher temperatures and acts in the direction of breaking the existing molecular order. At the point of clarification, however, the induced anisotropy is eliminated and the system turns into an isotropic liquid. The nematogenic molecules used as solvents in their mesomorphic interval "carry away" the molecules of the dissolved substance, causing partial orientation of the latter. The effect of dissolution is implemented by heating the mixture above the point of clarification and subsequently cooling it in the cuvette. This cuvette contains a layer of 0.002 to 0.004 cm thickness, and the KBr plates are preliminarily "roughened" by making "channels" in one direction only with appropriate tissue or fine glass paper, for example, C800. The induced monoaxial orientation is implemented in this direction. The optimum concentration of the dissolved substance is 5%. Regardless of the huge variety of compounds with nematic properties, the scope of solvents used in the IR-LD spectral analysis is limited by the requirement for optimum transparency in the middle IR region in order to prevent any "masking" of absorption bands for the substance studied. Particularly appropriate in this respect are the nematics of the bicyclohexyl type, especially 4-cyano-4′-alkylbicyclohexyl-type eutectic mixture ZLI-1695® (Merck Darmstadt, Germany) (Figure 1.10).

The liquid crystals of the aforementioned type have an IR spectrum, which resembles that of Nujol and poly(ethylene), and the band, corresponding to the CH-rocking vibrations within 720 to 750 cm$^{-1}$, is practically not observed (see Figure 1.11). The presence of the end cyano group at 2235 cm$^{-1}$ makes the IR-LD spectral analysis of compounds with a triple bond in their molecular structure difficult. On the other hand, the intensity of the cyano group in the differential spectrum is an excellent indicator for the degree of orientation of the sample. Compared to the other liquid crystals of this class, ZLI-1695 has the advantage of possessing a mesophase within a wider temperature interval, including the room temperature region, and the use of additional thermostating devices is not necessary.

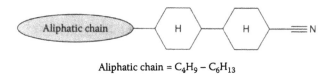

Aliphatic chain = $C_4H_9 - C_6H_{13}$

**FIGURE 1.10**  Structure of the nematic liquid crystal of the alkylbicyclohexyl type.

The principal advantage of the methods of orientation in nematic solvents, compared to the technique in a poly(ethylene) matrix, is the facile and prompt preparation
of samples as well as their stability on storage. Moreover, the use of cuvettes with fixed
thickness of the layer makes it possible to compensate for the inherent absorption bands
of the liquid crystal, except for those at 2800 to 3000 $cm^{-1}$. The manipulation, however,
is not always successful, since there is the possibility for both the solution and pure nematic to have different degrees of orientation. The latter would, eventually, result in the
residual absorption, observed in the reduced spectra and caused by solvent molecules.

Another drawback of the method of orientation in nematic solvents is the possibility of interactions between the dissolved substance and the solvent, which is practically absent when the inert poly(ethylene) is employed. Such an effect can be reduced
when using the relatively nonpolar solvents of cyano-alkyl-bicyclohexyl type. It has
been shown, however, that the presence of ZLI-1695 affects the structure of some
compounds such as 4'-cyano-phenylthiolbenzoate [28]. An additional disadvantage
of these nematics of low polarity is associated with their limited solvating capability
with respect to polar guest compounds. The influence of the liquid crystal medium
on the IR-spectral parameters has been discussed by Korte and Schrader [67].

### 1.3.2  ORIENTATION AS SUSPENSION IN LIQUID CRYSTALS

Suspension in liquid crystals is an easy-to-perform technique similar to the preparation of samples in Nujol; however, the nematic substance, such as a nematic liquid
crystal, which is mesomorphic at room temperature, is used as a medium (see later).
The suspension of low viscosity containing approximately 5 mg of the substance
studied is deposited between preliminarily roughened KBr plates as described earlier. Consequently, the obtained thin layer is additionally powdered into the plates in
the direction of their roughened channels. Satisfactory and, in some cases, even good
orientation of samples can be achieved by this manipulation.

Regardless that the problem of the complicated interpretation of solid-sample
IR spectra remains unresolved, this method offers two essential advantages. The
analytical samples are easily prepared without preliminary treatment, except for
the fine powdering of the compound studied. The preparation is conducted at room
temperature, and thus, various substances can be oriented, regardless of their melting and decomposition temperatures and crystallization capability. IR-LD spectra
have been obtained with samples, prepared as suspensions in liquid crystals, and
important structural information about a number of organic compounds, their salts
or complexes, including substances of proven biological activity has been provided.
Additional information about tautomeric equilibriums and polymorphism has also
been obtained.

## 1.4   PHOTOMETRIZATION AND PROCESSING OF IR-LD SPECTRA: REDUCING DIFFERENTIAL PROCEDURE

It was pointed out that studies involving the linear polarization phenomena are implemented by a measurement and comparison of the intensities of the IR spectral bands at various angles of polarization, most frequently parallel ($IR_p$) and perpendicular ($IR_s$) with respect to the orienting director **n**. This qualitative approach is characteristic for the interpretation of data associated with earlier work on the IR-LD spectral analysis. This is mainly because the IR spectrophotometers from the previous generations registered the absorbed irradiation as a percentage of transparency. Any quantitative comparison was then only possible if the classical baseline method was utilized. This approach is not correct and it is even not applicable in the case of superimposing bands. However, the quantitative interpretation of spectra is of principal importance. It offers a totally different approach, since the logarithm of the intensity ratio for the $i$-band in both polarization states corresponds to the difference between the absorptions [6,7,17]:

$$\lg\frac{I_S}{I_P} = \lg\frac{\dfrac{I_S}{I_0}}{\dfrac{I_P}{I_0}} = \lg\frac{I_S}{I_0} - \lg\frac{I_P}{I_0} = \lg\frac{I_0}{I_P} - \lg\frac{I_0}{I_S} = A_P - A_S \tag{1.29}$$

This mode of registration is called *differential* polarization spectroscopy. For the dispersive IR spectrophotometers, it is implemented by placing the sample along the common optical pathway, where the measurement and reference beams, polarized at the angle of 90° with respect to each other, gather together after modulation of the signal. This is achieved by assembling two polarizing devices in the instrument. The direct IR-LD spectrum of the compound studied is thus obtained as ratio of the corresponding intensities $I_p^v / I_s^v$ or the difference between absorptions ($A_p^v - A_s^v$) if the latter can be calculated.

The method was first published by Zbinden [2] on the basis of a personal communication with Tink and Marrinan concerning the IR-LD analysis of extended polymers. Modification of the method for the circular dichroism was developed by Korte [67]. Called *common beam spectroscopy* (CBS), the application of the method was further expanded by Jordanov and Tsankov for the measurement of the four variations of infrared dichroism: linear and circular polarization, and linear and circular beam refraction [68–70].

The vast introduction of the more sophisticated Fourier transform infrared (FT-IR) spectrometers for spectral studies made the CBS method senseless in terms of its further application. In the FT-IR analyses, the linearly polarized measurements are conducted by sequential photometrization of a sample for both positions of the polarizer. The linear dichroism is differentially registered by subtraction of the two spectra, which is implemented promptly and easily by the modern sophisticated FT-IR spectrophotometers, equipped with computerized systems.

The *differential spectrum* obtained by the subtraction of the perpendicular ($IR_s$) from the parallel ($IR_p$) spectrum consists of both positive (+) and negative (−) peaks. These peaks correspond to bands, which are more intensive for parallel (+) and perpendicular (−) polarization. The angle $\theta$ between the corresponding moments of transition and the orientation director $\mathbf{n}$ (see Section 1.1) represents the determining factor in this respect. The analytical expression of this dependence is [52]:

$$A_P^i - A_S^i = \frac{1}{2} S A_i \left(3\cos^2 \theta_i - 1\right) \qquad (1.30)$$

where $S$ represents the parameter of order, $A_i$ is the absorption of the $i$-band in the nonpolarized spectrum, and $\theta_i$ is the angle between the moment of transition of the $i$-band and director $\mathbf{n}$. For $0° \leq \theta_i < 54.7°$ and $(A_p - A_s) > 0$, the sign of this expression changes at $54.7° < \theta_i \leq 90°$. The value of the magic angle $\theta_i = 54.7°$ corresponds to $3\cos\theta_i = 1$ and nullifies Equation 1.23. The band is then eliminated in the differential IR-LD spectrum.

The parallel and perpendicular IR-LD spectra of the liquid crystal ZLI-1695 are shown in Figure 1.11. As outlined later, the strongly pronounced anisometric shape of the molecules of the bicyclohexyl type (Figure 1.10) determines their very good orientation along the long molecular axis and the high value of the parameter of order $S$. For such an anisotropic structure, the moment of transition of the $C \equiv N$ stretching vibration directed along the length of the nitrile group almost coincides with the direction of orientation $\mathbf{n}$ (Figure 1.11). The corresponding band at 2235 cm$^{-1}$ has strong intensity in the parallel, a weak one in the perpendicular, and positive intensity in the so-called difference spectrum (obtained by subtracting perpendicular spectrum from parallel). For the stretching- ($\nu^s$) and scissor-bending ($\delta$) vibration of the methylene groups in the cyclohexyl nucleus, the moments of transition are directed along the half bisector of the H–C–H angle. These moments "pin through" the double cone, associated with the angle $\theta = 54.7°$ (Figure 1.11), and the bands within 3000 to 2800 cm$^{-1}$ as well as those at 1450 cm$^{-1}$ have a negative sign.

The differential approach facilitates significantly the interpretation of IR-LD spectral data, since it provides for the easy identification and characterization of bands, according to their symmetry origin. A typical example, a fragment from the IR-LD analysis of S-phenyl-4'-cyanothiolbenzoate, is presented [28]. The anisometric geometry of this compound (Figure 1.12) provides an excellent possibility of orientation in solution as well as in solid state. The differential IR-LD spectrum of its polycrystalline film is shown in Figure 1.13. This spectrum contains the positive band corresponding to the $C \equiv N$-stretching vibration at 2229 cm$^{-1}$, which suggests a significant degree of orientation along the long molecular axis (Figure 1.13). Under these conditions, the bands for the $p$-substituted benzene nucleus, with moments of transition coinciding with the molecular axis and determined by the two substituted carbon atoms of the ring ($A_1$-type vibrations), should have positive signs. At the same time, the $B_1$-type vibrations, perpendicular with respect to the director $\mathbf{n}$ and the plane of the aromatic ring (Figure 1.12), will be "negatively" directed. The stretching

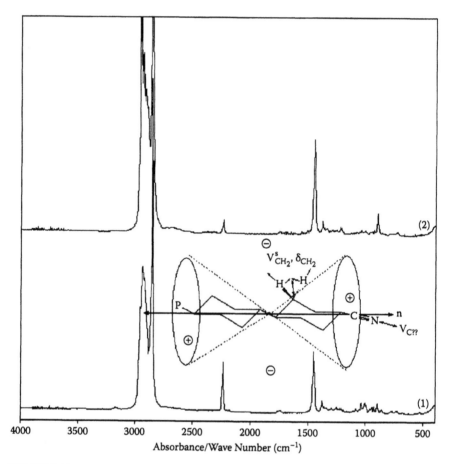

**FIGURE 1.11**   (1) Parallel and (2) perpendicular IR-LD spectra of ZLI-1695.

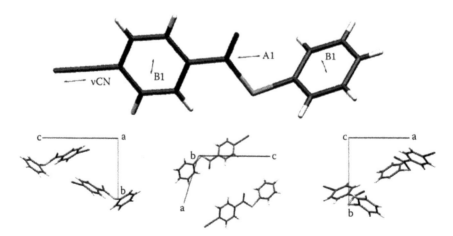

**FIGURE 1.12**   Molecular structure of S-phenyl-4′-cyanothiolbenzoate according to Ivanova [71], and the view along the **a**, **b**, and **c** axes.

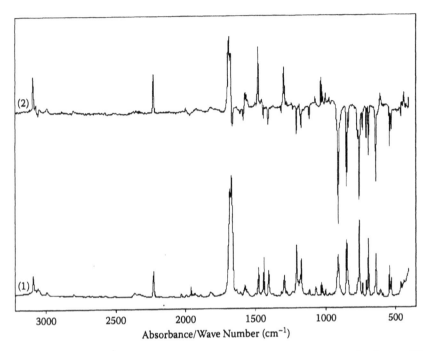

**FIGURE 1.13** (1) Nonpolarized IR and (2) differential IR-LD spectrum of polycrystalline film of S-phenyl-4'-cyanothiolbenzoate.

vibration Ar-H- as well as the skeletal one (C=C$^{Ar}$) for the *p*-substituted nucleus at 3093 and 1479 cm$^{-1}$, respectively, belong to the first type. The out-of-plane bending vibrations in the lower frequency interval, corresponding to the series of negative bands between 1000 and 600 cm$^{-1}$, and more particularly the γ-CH-absorption maximum at 846 cm$^{-1}$ for the *p*-substituted benzene ring, belong to the second (B$_1$) type of vibration.

The *stepwise reduction* method was proposed by Thulstrup and Eggers [5] as an efficient development of the differential approach. Initially, this method was employed for the determination of the polarization of electron transitions. Later, Michl and Radziszewski [72] adapted it for processing IR-LD spectra of compounds, oriented within a poly(ethylene) matrix. Generalized analysis of both the theory and application of this method was made by Jordanov [16,19].

The basics of the *differential reduction* method can be outlined as follows: The automated procedure for the registration of a differential spectrum is conducted by the introduction of a coefficient of unity into the processing program "subtraction." *Spectral subtract* performs an automatic subtraction of data files. The program for spectra interpretation can automatically determine the best parameters for any two given files. It can also perform subtraction using user-defined parameters, using a subtraction function, based on an algorithm called "dewiggle" as described in Banerjee [73,74]. By interactively varying the subtraction factor and looking at the results, the user can determine when the subtrahend has been adequately "removed" from the sample data. This is a commonly used function throughout all spectroscopy software packages. Thus the direct differences from the spectral curves for parallel and perpendicular polarization are registered as positive and negative absorption maxima. If, however, the interactive operation is localized on a certain band and its nullification is followed on the computer screen, the value of the corresponding coefficient is different from unity ($c \neq 1$; the specific case with $c = 1$ shows that this band is not present in the differential spectrum). The following equation is therefore applied in this case:

$$A^i_p - cA^i_s = 0 \qquad (1.31)$$

or

$$c = \frac{A^i_P}{A^i_S} \qquad (1.32)$$

Equations 1.32 and 1.17 are identical, which means that, at the point of elimination, the parameter $c$ *coincides with the dichroic ratio of the i-band*. Consequently, in the reduced IR-LD spectrum obtained, all absorption maxima, generated by moments of transition that are parallel to those of the *i*-vibration are not present. This conclusion is mostly valid for the vibrations belonging to a given symmetry class and also for random colinearity of the moments of transition caused by the molecular or super molecular geometry (see Section 2.5 in Chapter 2). By varying the spectral subtraction factor some bands of the difference spectrum can be eliminated. At this moment the factor becomes equal to the dichroic ratio in the vanished bands. The orientation parameters of a given vibrational transition moment $\vec{M}_u$ of the $u$th molecular direction is obtained from the dichroic ratio as [17]:

$$K_u = R_u / (R_u + 2) \qquad (1.33)$$

In formal vibrations, proceeding along three mutually orthogonal molecular axes ($u = x,y,z$) induced by the molecular symmetry, then the following relationship for $K_u$ is valid:

$$K_x + K_y + K_z = 1 \qquad (1.34)$$

After removing the bands belonging to a direction, say, w, the reduced spectrum remains as a linear combination of the other two absorbances, $A_u$ and $A_v$:

$$\Delta_{uv} = [K_u + 1/2R_w(K_u - 1)]A_u + [K_v + 1/2R_w(K_v - 1)]A_v,$$
$$\text{where } u,v,w = x,y,z \quad \text{for} \quad x \neq y \neq z \qquad (1.35)$$

The procedure of differential reduction strongly enhances the effectiveness of the IR-LD spectral analysis and represents an excellent approach toward the modern application of the method for structural characterization of chemical compounds. It is also of significant importance for the interpretation of IR spectra. Besides providing direct information for the classification of vibrations, according to their symmetry class, this reducing technique contributes greatly to the identification of bands obtained by random degeneration, for example, in systems oriented in NLC solution by Jordanov [24,75]. The application of the methods for the identification of crystals—Fermi resonance (FR), Bethe splitting (BS; also called the Davydov splitting [DS] effect), Fermi–Davydov (FD) resonance effects, and Evans's hole (EH) effect—have been developed further (see below for part of the analysis of crystals).

The essential advantage of the reducing method is that the results obtained do not depend on the orientation director $\mathbf{n}$ of the sample. According to Equation 1.32, the nullification of the $i$-band is implemented at $A^i_p = A^i_s$, which corresponds to $\theta = 54.7°$. The geometric equivalent of this numerical procedure represents rotation of the corresponding moment(s) of transition with an angle equaling the magic one. In other words, depending on the direction of macro orientation, the parameter $c$ (Equation 1.32) will have a different value but will provide an identical result with respect to the elimination of bands of the same dichroic ratio. This conclusion is important for the IR-LD spectral analysis of solid samples, for which the macro direction of anisotropy is determined by the applied method for orientation and by the effect of the orienting medium.

An illustration of the application of the reducing approach can be considered for the already discussed case of S-phenyl-4'-cyanothiolbenzoate. As mentioned earlier, the bands at 2229 cm$^{-1}$ ($\nu$(CN)), 3093 cm$^{-1}$ ($\nu$(C-H$^{Ar}$)), and 1479 cm$^{-1}$ ($\nu$(C=C$^{Ar}$)) have high intensities and positive signs in the differential spectrum. This leads to the conclusion that the director $\mathbf{n}$ is located along the long molecular axis, which, on the other hand, coincides with the indicated moments of transition of the bands (Figure 1.12). With this prerequisite, the nullification of any of the three absorption maximums should result in the elimination of the remaining two, which is actually observed. For the $\nu^{Ar}_{C-H}$ and $\nu^{Ar}_{C=C}$ bands, the result is automatic, since the corresponding vibrations belong to the same ($A_1$) class of symmetry. The parallel elimination of the $\nu_{C\equiv N}$ band at 2229 cm$^{-1}$ represents, however, an example for the coincidence of the moments of transition with respect to the colinearity. In this case, it is determined by the effect of sp respectively sp$^2$ conjugation between the triple bond and benzene ring. This, in turn, predetermines the orientation of the cyano group along the molecular axis, defined by the $p$-substituted aromatic carbon atoms, located oppositely.

## 1.5 EFFECTS IN THE INFRARED SPECTRA OF CRYSTALS

The investigation of the vibrational spectra and, in particular, the IR spectra of hydrogen-bonded systems in the solid state is an important topic in physics, chemistry, and biology. However, these spectra are complicated as a result of significant anharmonic effects. Anharmonicity in the high frequency intermolecular vibrations in crystals has been studied over many years. Two aspects complicate this problem. The first of these is Fermi resonance (FR), which occurs between a fundamental vibration and a combination (or overtone) of the molecule. The second is splitting, which arises from intermolecular interaction and transmission of vibrational excitation from one molecule to another in a crystal. As a result, one nondegenerate vibration of the free molecule may be split in the crystal into a multiplet of bands [76]. In 1931 Frenkel [77] postulated the nature of the excitation wave in the crystals. In 1942 Freed [78] showed that even before Bethe [76], employing group theory laid the groundwork for the study of energy states in the crystal. Later, Davydov's [79,80] work, especially concerning monoclinic crystals, must also be mentioned. The number of components is equal to the number of molecules in the crystal unit cell and the polarization of these bands can be different (the effect was named Bethe splitting [BS] [81,82] or in some papers Davydov splitting [DS]). The problem of combined Fermi–Davydov (FD) resonance in crystals was considered by Lisitsa [83–88]. The problem of anharmonicity and intermolecular (inter-H-bond) interaction was first formulated by Marechal and Witkowski [89] in the case of a dimer where interaction between the two hydrogen bonds of the dimer was included. Later, Witkowski and Marechal [90] included the FR between the stretching mode and the overtone of the bending mode, and Wojcik [91] added to the FR the interaction between the hydrogen bonds of the dimer (a Bethe-type interaction). In these calculations, interaction with other subsystems was neglected and spectral densities appear as $d$ functions (similar to the results obtained by Lisitsa and Yaremko [86] and Valakh [87] for crystals). A modern approach to the study of the dimer hydrogen bond vibrations using the Green function method was recently proposed in the publications of Chamma and Henri-Rousseau [92] and Belharaya et al. [93], where the effect of the surroundings resulting in damping (direct or indirect relaxation) of the fast vibration was included. The problem of hydrogen bond vibrations and anharmonic effects in liquids has been discussed by Bratos [94,95] and Ratajczak [96,97]. For crystals, the similar damping effect was analyzed by including weak anharmonic interaction of the lattice vibrations with the intramolecular fast mode [87,88]. However, in the case of hydrogen-bonded crystals, the problem of FR and FD effects seems to be more complex because of strong anharmonic interaction of hydrogen-bond vibrations with the lattice phonons. This problem was studied using the Green function method [96,97] and damping of high frequency vibrations arising from anharmonic interaction was included as a parameter, similar to the approach used by Vogt [13], Valakh [87], and Lisitsa and Yaremko [88]; it was calculated as the self-energy part of the Green function. It should be noted also that for crystals having several identical, but differently oriented hydrogen bonds (molecular dipoles) in the unit cell, additional considerations apply [98].

The discussion about the crystal field splitting effect developed by Bethe [76] and later called Bethe splitting [79] started with the problem of crystal field splitting at levels that are degenerated in the free molecule. The crystal field splitting may also be caused by transitions that are forbidden in the vapor. Another effect of crystallization is implicit in the work of Frenkel [77] and Witkowski [99] on the theory of excitation and was later treated fully for monoclinic crystals by Davydov [79,80]. In 1931, Frenkel [77] postulated that "since the resonance conditions for the frequencies and wave-lengths are both unsharp, the excitation state of the crystal inducing the absorption of the energy must be represented not by one defined expiation wave, but by a superposition of a number of such waves with approximately equal lengths and frequencies or as a group of excitation states."

Wojcik [91] developed a theory that is applicable to weak and medium-strength hydrogen bonds. The basic physical mechanism responsible for the energy and intensity distribution is the coupling between the high-frequency hydrogen-stretching vibration and the low-frequency hydrogen bond O...O stretching vibration, which in a case of interacting hydrogen bonds is modified by resonance interactions. Though the theory has been formulated [99,100], numerical calculations are only possible for cases of a small number of interacting hydrogen bonds in symmetrical systems such as dimers and crystals. Formally, the approach is the vibrational analogue of vibronic coupling, like the pseudo Jahn-Teller effect in the electronic spectra of dimers [101] or molecular crystals [102].

Another effect frequently recalled in discussions about the origin of the band shape of hydrogen bonds (especially strong ones), is the FR [103,104] between the hydrogen-stretching vibration and overtones or combination vibrations of hydrogen-bonded complexes, most probably involving the hydrogen bending or the stretching of a bond nearest to the O-H group, such as C=O. Unfortunately, nobody has succeeded, until now, in using FR theory to obtain a quantitative comparison between experimental and theoretical spectra for real systems. Moreover, the nature of the "pseudo maxima" in the IR spectra of the strongly hydrogen-bonded systems has been studied, taking into account the possible role of FR. It has been shown that such pseudo maxima do not correspond to fundamental vibrations [105]. In this respect, in the earlier Evans's work (on the EH effect) [106] it was been shown that when a fundamental vibration, which gives rise to a broad IR band, overlaps a sharp fundamental band of the same symmetry this can lead to a *transmission window* in the broad band at the frequency of the sharp fundamental. Possible mechanisms have been present [103,104]: (1) coupling of the low- and high-frequency motions modified by resonance interactions and (2) Fermi resonance between fundamental and overtone (or combination) states produce the infrared band structure, their relative contributions depending on the strength and nature of the hydrogen-bonded system. We should underline that in the current literature there is still some controversy about the interpretation of the complex vibrational spectra in the range of a stretching vibration ($\nu_{XH}$) band. Several mechanisms have been postulated, such as the coupling between the $\nu_{XH}$ and low-frequency vibrations, or lattice phonons, FR, or tunnelling vibrational splitting. It is difficult to study these effects by conventional infrared spectroscopic tools, which is one of the great limitations of the method for studying the crystalline compounds. However, in all of the aforementioned fundamental research, it has been demonstrated that the components of the multicomponent spectroscopic patterns have different polarizations, depending on the origin. For these reasons, the possibilities of IR-LD spectroscopy can be used for experimental assignment of the phenomena in crystalline compounds. In this respect we have been using an orientation method involving colloid suspensions *in a nematic liquid crystal* host in our research.

The background of the theory of the observed IR-spectroscopic effect in crystals can be summarized in the following way. If in the harmonic approximation an isolated impurity molecule could be characterized by two adjacent levels ($p$ and $q$), then in an anharmonic condition we get a new pair of more remote levels ($f$ and $g$). As experiments show [80], in vibrational spectra the value of FR for isolated molecules is often comparable with the value of the DS for the molecular lattice (about several tens of cm$^{-1}$). Therefore, in zero$^{th}$ approximation one can formulate expressions for the wave function of an excited impurity crystal. Using these expressions, Lisitsa and

Yaremko [107] described that the separation between the $f$ and $g$ levels in an isolated molecule is determined by the value of the FR interaction. The splitting of the non-degenerate $f$ level (or $g$ level) associated with different orientations of the molecules within the unit cell is determined by the value of the intermolecular interaction. Both boundary cases of FR or DS effects could be described.

1. An overlapping of the wave functions corresponding to $f$ and $g$ states is much smaller than that of the wave functions corresponding to the Bethe components arising as a result of splitting of $f$ and $g$ states. Then, $f$ and $g$ levels resulting from $p$ and $q$ levels can be written separately because of the influence of anharmonicity. When exact resonance is present

$$d^f = \frac{1}{\sqrt{2}}(d^p - d^q)$$

$$d^g = \frac{1}{\sqrt{2}}(d^p - d^q)$$

(1.36)

where $d^p$ and $d^q$ are dipole moments of transition of molecules from basic state into $p$ and $q$ states. In our case the absolute values of $d^p > d^q$ are therefore $d^f \cong d^g$. Then it follows from Equation 1.36 that for values of $\theta^f$ and $\theta^g$ approaching one another the dipole moments $d^f(+)$ and $d^g(+)$ will exhibit almost the same polarization as $d^f(-)$ and $d^g(-)$, whereas $d^f(+)$ and $d^f(-)$ will be normal to each other. Thus two couples of mutually orthogonal bands will be present in the absorption spectrum. If some transitions are forbidden, the spectrum will have fewer components.

The ratio of intensities of bands with the same polarization will be

$$f_{\pm} = \frac{1 \pm \sin\theta^f \cos\nu}{1 \pm \sin\theta^g \cos\nu}$$

(1.37)

A couple of bands corresponding to the positive sign will be polarized normally to that corresponding to the negative sign.

The last formula shows that for $\theta^f \pm \theta^g$ either $f^+ > 1$ or $f^- > 1$. This means that for one of the polarizations a component resulting from a harmonic of higher order will have greater intensity than another component.

2. In the opposite case, the overlapping of the wave functions corresponding to the Fermi components will be much stronger than that for the Bethe's components. The corresponding equations are

$$d^{\alpha}(+)=\sqrt{N}(\cos\frac{\theta^{\alpha}}{2}d^f_{\alpha}+\sin\frac{\theta^{\alpha}}{2}d^g_{\alpha})$$

$$d^{\alpha}(-)=\sqrt{N}(-\sin\frac{\theta^{\alpha}}{2}d^f_{\alpha}+\cos\frac{\theta^{\alpha}}{2}d^g_{\alpha})$$

(1.38)

It is seen that $d^{\alpha}(+)$ and $d^{\alpha}(-)$ have the same polarization ($d^f_{\alpha} \cong d^g_{\alpha}$) for any value of $\theta$. They can be different in absolute values with the obtained values corresponding to molecules of another orientation. Polarization of the bands in case $\beta$ will differ by an angle $v$ from polarization of the bands $\alpha$.

The ratio of intensities for bands with the same polarization will be

$$f^{\alpha,\beta}=\frac{\sqrt{1+tg^2\theta^{\alpha,\beta}}+tg\theta^{\alpha,\beta}}{\sqrt{1+tg^2\theta^{\alpha,\beta}}-tg\theta^{\alpha,\beta}}$$

(1.39)

where a choice of indexes $\alpha$ or $\beta$ depends upon the orientation of the molecules for which the ratio of band intensities is considered.

# 2 The Background of the Orientation Method of Colloid Suspensions in a Nematic Host

## 2.1 ORIENTATION PROCEDURE

The orientation method *colloid suspensions in a nematic liquid crystal* (CS-NLC) [32–35] shows that a partial orientation (15% to 20%) of suspended particles, adequate for the recording of reasonable linearly polarized infrared (IR) spectra, is achieved when 5% ± 1% by weight of the given solid compound with particle size within the limits of 0.3 to 0.9 μm is mixed with a nematic liquid crystal substance suitable for IR spectroscopy and the slightly viscous suspension obtained is phase-pressed between two potassium bromide (KBr) plates (Figure 2.1). The latter are roughened in one direction with fine sandpaper prior to use (C800, size 5 μm). Then, the KBr plates and pressed suspension are moved repeatedly with 3 μm/s for 100 times. The optimal cell thickness is 100 μm (Figure 2.2). If mathematical procedures, including deconvolution, curve-fitting second derivative analysis for interpretation of nonpolarized IR spectra, and the reducing-difference procedure (RDP) for analysis of polarized IR spectra are applied, it is possible to perform detailed IR characteristic band assignment and structural elucidation of the embedded compounds in the solid state. These possibilities are independent of the physical properties of the suspended samples, such as their melting point, crystalline or amorphous state, and the quality of the single crystals or polycrystalline samples. This means that the method permits an investigation of amorphous solids, high-temperature melted organic and inorganic compounds, metal complexes, and glasses cannot be successfully examined by the known orientation methods applied in linear-dichroic infrared (IR-LD) spectroscopy.

The development of the presented orientation technique during the last 4 to 5 years has included the elucidation of fundamental questions concerning the morphology of the suspended particles, the particle size, and the influence of the physicochemical properties of the liquid crystal medium on the degree of orientation of suspended particles. Other questions of interest were the velocity of the shearing of the suspension, the degree of the roughening of the KBr plates and their effects on the degree of orientation, the influence of the space group on the orientation parameter, the nature and balance of the forces acting on the suspended particles and their degree of orientation, and last but not least, the mathematical model that best describes our heterogenic system. Modern statistical approaches have also been applied to

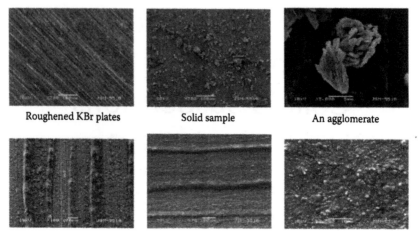

| Roughened KBr plates | Solid sample | An agglomerate |

Oriented colloidal suspension on a roughened KBr plate        Distortion of the orientation after 2h

**FIGURE 2.1** Electron microscopic data. (Koleva, B., T. Kolev, V. Simeonov, T. Spassov, and M. Spiteller, 2008, *J. Incl. Phen.* 61:319–333. With permission.)

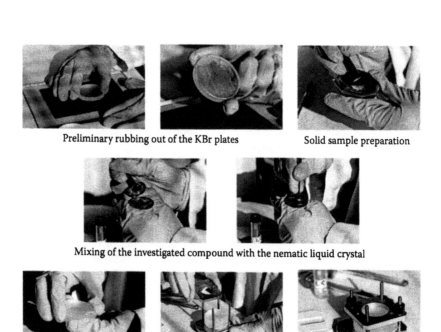

Preliminary rubbing out of the KBr plates          Solid sample preparation

Mixing of the investigated compound with the nematic liquid crystal

Placement of the suspension between the KBr plates and then in the holder

**FIGURE 2.2** Preparation of the oriented colloidal suspension in a nematic liquid crystal. **(See color insert.)**

estimate the impact of the experimental parameters, for example, size, velocity or thickness, on the IR signal. An experimental design involving full factorial design on two levels of variation of the input factors was included as well [34,35]. Validation of the RDP for accuracy and precision was established. The questions concerning the limits of detection and smoothing procedures for IR spectral analysis based on Savitzky–Golay or Fourier methods are studied as well. The mean values and relative standard deviations for the peak position ($v_i$) and integral absorbance ($A_i$) by this data processing approach were examined using the Student's $t$-test [33]. The validation of the method included repeatability, the influence of the liquid crystal medium on the peak positions and integral absorbencies of the guest molecule, determination of the optimal experimental conditions, and experimental design. In this respect, the impact of four different input experimental factors were studied in a qualitative way as the number of scans, the rubbing out of the KBr pellets, the amount of the compound studied included in the liquid crystal medium, and the ratio of Lorentzian and Gaussian peak functions in the curve-fitting procedure [33–35].

It was found that for the oriented solid samples as colloid suspensions in an NLC, liquid crystal mesomorphic media such as ZLI 1695, MLC-6815®, and ZLI-1538® (Merck, Germany) are suitable. Like in the well-known technique for investigation of dissolved compounds in NLC solution [14–31], the selection of the orientation medium is based on the premise that their self-absorption in the middle IR spectroscopic region must be minimal (Figure 1.11). 4-cyano-4′-bicyclohexyl liquid crystal is a suitable host medium for orientation, since its weak IR spectrum permits recording of the guest-compound IR bands in the full 4000 to 400 cm$^{-1}$ range. Apart from the regions 2800 to 3000 cm$^{-1}$, and near 2235 cm$^{-1}$, 1457 cm$^{-1}$, and 896 cm$^{-1}$ in ZLI-1695 (Figure 1.11) and ZLI-1538, the rest of the region is available. In the case of MLC-6815, the presence of the ester groups narrows the IR spectroscopic range possible for analysis obliterating the regions close to 1723 cm$^{-1}$ and from 1447 to 896 cm$^{-1}$ (Figure 2.3). A mesomorphic interval including room temperature (25°C) is another important requirement for an orientation medium. ZLI-1695 and MLC-6815 both have a mesomorphic interval that includes 25°C. However, the S-N transition temperature of ZLI-1538 is 54°C, so the suspension must be prepared and maintained in a thermostat. Other advantages of the selected liquid crystals are their physico-chemical properties:

ZLI-1695—Clearing point of 72°C, viscosity 62 mm²/s at 20°C, dielectric anisotropy 4.2 at 20°C and 1 kHz, optical anisotropy 0.0600 at 20°C and 589 nm

MLC-6815—Clearing point of 67°C, dielectric anisotropy 8.1 at 20°C and 1 kHz, optical anisotropy 0.0517 at 20°C and 589 nm

ZLI-1538—Clearing point of 79°C, viscosity 68 mm²/s at 20°C, dielectric anisotropy 3.8 at 20°C and 1 kHz, optical anisotropy 0.0500 at 20°C and 589 nm

It is possible to orientate polar solid compounds in ZLI-1695 and ZLI-1538 and non-polar compounds in MLC-6815. The isolated nitrile stretching IR band at about 2235 cm$^{-1}$ also serves as an orientation indicator.

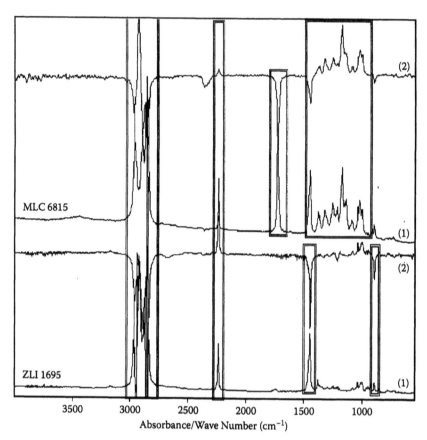

**FIGURE 2.3**   (1) Nonpolarized IR and (2) differential IR-LD spectra of the liquid crystals ZLI-1695 and MLC-6815.

The moving of the KBr plates against each other is used to promote additional orientation of the liquid crystal suspension (Figure 2.1). The process in the CS-NLC system cannot be described with Couette flow, using the continuum Leslie–Ericksen theory [108], where the orientation of a pure liquid crystal is promoted by moving the plates. The colloidal suspensions are different systems because of their specific properties, that is the long-range deformation field created by the particles in the LC, as a result of the director anchoring of the particle surface. In the dipole configuration the particle is accompanied by a topological point defect, whereas in the Saturn-ring configuration the particle is surrounded by a −1/2 disclination ring at the equator [38,109]. The orientation defects observed near colloidal particles in an NLC have been studied intensively [41,42,48,110–117]. It is well known that when solid particles are introduced into a nematic medium, the nematic molecules prefer to orient at a certain angle on the particle surface, for example, perpendicularly in so-called homeotropic anchoring, which is observed on KBr plates as well. We postulate that the threshold for these processes will be lowered by a significant amount by first roughening the KBr plates and then moving the system in a single direction

**n**. In West et al. [111] it was noted that the anchoring direction of particle surfaces would come into conflict with far field orientation. It is believed that the resulting defects play a critical role in the interactions between the particles and the novel microstructures formed [118]. These assumptions are more important in our case where the suspended particles are far from being ideal spheres. The Frank theory or linear elastic theory, based on the molecule orientation distribution being uniaxial about the director **n**, breaks down in such cases [118]. Some of the phenomena discussed remain negligible when the KBr plates are moved repeatedly backward and forward in a single direction (100 times) [35] with a velocity of 3 μm/s causing orientation of the suspended particles (Figure 2.1), which also promotes orientation. Hence, roughening the KBr plates in advance is advantageous. Effective orientation of the sample is achieved when the separation between the two KBr plates, that is, cell thickness, is 100 μm. Measurements must be carried out within 2 h of preparation of the system because there is significant breakdown in the degree of orientation after this period (Figure 2.2).

The detail investigations of model organic systems oriented in liquid crystal hosts agree extremely well with the dependences established in the critical review by West and coworkers [117] on the drag on colloidal particles by a moving nematic-isotropic interface. The different forces act in the colloidal suspension and several mechanisms affect the total drag force acting on a particle. The surface tension coefficient might differ and an additional pressure ($P$) might be caused by the curvature. The equation applicable is $P = f(\sigma, R)$ and $\Delta\sigma = \sigma_N - \sigma_I$, where $R$ is the radius of the particle and $\sigma$ is surface tension coefficient. This pressure contributes to the total drag force as $F_\sigma$, with the amplitude growing linearly with the droplet radius $d$, the distance from the particle center to the interface. $\Delta\sigma$ depends on the surface treatment of the particles and is unknown for the CS-NLC system and would be difficult to measure. The order of magnitude can be estimated from the change of the surface tension coefficient of each of the organic particle-liquid crystal interfaces and the values of $\Delta\sigma$ lie within the range $10^{-2}$ to $10^{-3}$ dyn cm$^{-2}$. The particle creates long-range distortions of the director **n** in a nematic phase. In our case to minimize elastic distortion energy, the nematic phase attempts to order particles along the axis parallel to **n**. The elastic forces have two origins: as a result of the director deformations in the bulk nematic phase and as a result of the director being anchored at the particle surface. An estimate of these contributions can be made by dimensional analysis. For the surface contribution, the only combination that has the dimensions of force is $WR$, where $W$ is the anchoring coefficient. Therefore, the surface contribution to the drag force $F_s$ is proportional to $WR$. One can have two different situations for the bulk contribution. For *weak* anchoring, $WR/K < 1$, the bulk contribution is proportional to the squared characteristic deviation of the director, $\beta_0 \sim WR/K$, where $K$ is the elastic constant. Now, $W^2R^2/K$ has the dimension of force, yielding $F_b$. In the case of strong anchoring, $WR/K > 1$, the anchoring does not enter the elastic contribution. The liquid crystals used here possess typical values of the anchoring energy, $W \sim 10^{-3}$–$10^{-4}$ dyn.cm$^{-1}$. $WR/K$ is $< 1$ for silica particles and $WR/K$ is $\sim 1$ for polymer particles, respectively. Hence, for silica particles, we have the weak anchorage regime, while polymer particles provide strong anchorage of the director. In the CS-NLC case, we must take both boundaries into account. We also note that the effective radius increases when particles agglomerate and we then have a strong anchorage regime even for small particles. A friction drag contribution, given by the Stokes formula, $F_\eta$, results in the $F_{drag}$ [117].

Solution of Newton's equations of $F_{drag}$ as a force completes the description of the particle dynamics. It is clear, however, that small heavy particles cannot be moved by the interface (see electron microscopy data). The maximum radius has been estimated from the conservation of linear momentum. To capture a particle of mass ($m$), the interface has to transfer to it a linear momentum ($mv$). If we assume that the particle does not move (or it moves much more slowly than the interface, which is valid for massive particles), then the total linear momentum transferred to the particle is

$$mv = \int_{t_1}^{t_2} F_{drag}\, dt = \frac{1}{\upsilon} \int_{-R}^{R} F_{drag}(x)\, dx \qquad (2.1)$$

If it is assumed that the interface touches the particle at time $t_1$ and leaves it at time $t_2$, $x = vt$. Substituting $F_{drag}$:

$$R_{max} = \frac{\frac{8}{3}\pi\Delta\sigma + \delta_s W - 6\pi\eta\Delta r}{\frac{4}{3}\pi\rho\upsilon^2 - \delta_b W^2 / K} \qquad (2.2)$$

The $\delta_i$ are geometry parameters, $\rho$ is the density of the particle, and $\Delta r$ is the final displacement of the particle resulting from the drag force. Several important conclusions can be drawn. First, if the particle is too large, the moving interface is not able to transfer sufficient linear momentum to it. Only particles with $R < R_{max}(v,W,s)$ will be captured by the interface. Only a slowly moving interface is able to capture the particles. The estimate of this velocity is given by the zero of the denominator of $v \sim W/(K\rho)^{-1/2} \sim 1$ μm/s. This is of the order of the limiting velocity for the cellular structure that we observed in our experiments: if the interface moves more slowly, then stripes appear, otherwise the cellular structure forms.

The main conclusion is that $R_{max}$ is a function of the material parameters, that is, it can be effectively controlled, for example, by changing the surface treatment of the particles, anchoring energy $W$. An increase in the anchoring energy leads to an increase of $R_{max}$. Moreover, sufficiently strong anchoring favors formation of a defect near the particle, contributing to an even higher energetic barrier created by elastic forces. On the other hand, if the particle is captured by the interface, the elastic force scales as $R^2$, and the opposing viscous drag scales as $R$. Therefore, there is a minimum radius, $R_{min}$, starting from which particles will be dragged by the interface. If the particle is dragged by the interface at a constant velocity, then $F_{drag} = 0$, yielding

$$R_{min} = \frac{6\pi\eta\upsilon - 2\pi\Delta\sigma - \gamma_s W}{\gamma_b W^2 / K} \qquad (2.3)$$

where $\gamma_i = g_i(0)$ are constants. To be moved by the interface, the particles have to be large enough. The elastic forces can only overcome viscous drag when this is the case. Substituting values typical for the liquid crystals employed and using the slowest cooling rate, we obtain $R_{min}$ ~0.08–0.15 µm, which agrees with the degree of orientation of suspended particles obtained experimentally. To explain the formation of the striped structure, we note that, in practice, the particles we are using aggregate into clusters, which are 1.0 to 10 µm in size. As has been noted [117], as an aggregate moves it captures more and more particles, growing in size. The anchoring parameter $WR/K$ also increases and we switch from weak anchorage to the strong anchorage regime. The bulk elastic contribution is then proportional to the elastic constant $K$ and the elastic force is no longer growing as $R^2$. Therefore, at some $R_c$, the friction drag overcomes the elastic contribution and the aggregate breaks through the interface. A stripe forms and the particles start to accumulate again. The condition $F_{drag} = 0$ gives the critical size of the aggregate

$$R_c = \frac{\gamma_b K}{6\pi\eta\upsilon - 2\pi\Delta\sigma} \tag{2.4}$$

which is about 1 µm for typical experimental values.

The calculation of the critical radius above which the particles cannot be captured by the moving interface predicts that the critical radius is sensitive to the viscous properties of the host liquid crystal for ZLI-1695, the value of the anchoring coefficient of the liquid crystal on the particle surface for MLC-6815, and the velocity of the moving interface for ZLI-1538 [35]. When the particle size of the suspended particles is less than 0.08 µm, reasonable orientation cannot be expected. As has been noted, the critical radius is sensitive to the viscous properties of the host liquid crystal, the value of the anchoring coefficient of the liquid crystal on the particle surface, and the velocity of the moving interface. To understand how the particles are moved by the nematic mesophase transition front, we used particulate organic compounds of different sizes. The influence of the morphology and of the crystalline particles on the degree of orientation in NLC suspension was studied using a selected set of compounds crystallizing in different polymorph modifications, space groups, and unit cell settings. The particle size was within the range 0.1 to 10.0 µm. In all cases the particles were dispersed at concentrations of 5% ± 1% by weight, which had been previously found to be optimum [35]. The suspension obtained was deposited between two KBr plates that had previously been roughened in one-direction plates. The cell thickness was varied between 50 and 150 µm in order to study the influence of layer thickness on the orientation of the sample. The direction of the stripes is parallel to the moving interface; the interface was moving from the left to the right of the cell in the geometry. The spatial period of the striped structure depended on the time and has been evaluated for the systems described. These results indicate that the particles are pushed by the moving nematic phase transition front.

According to the earlier equations, when the concentration of the suspended particles decreased the degree of orientation also decreased [46,119]. A reasonable explanation of this phenomenon could be found in the work of Araki

**FIGURE 2.4** Two types of defect formation around a pair of particles. A transient defect structure formed in the early stage becomes unstable and transforms into either, and a new type of (quasi-) stable configuration, in which a single disclination loop is shared by two neighboring particles.

and Tanaka [120] dealing with the colloidal aggregation in NLC. They studied many body interactions among colloidal particles suspended in an NLC, using a fluid particle dynamics method, which properly incorporates dynamical coupling among particles, nematic orientation, and flow field. Based on simulation results, we propose a new type of interparticle interaction in addition to the well-known quadrupolar interaction for particles accompanying Saturn-ring defects. This interaction is mediated by the defect of the nematic phase: upon nematic ordering, a closed disclination loop binds more than two particles to form a sheet-like dynamically arrested structure. The interaction depends upon the topology of disclination loop binding particles, which is determined by aggregation history (Figure 2.4) [121].

## 2.2    VALIDATION OF THE ORIENTATION PROCEDURE

The influence of the nature of orienting medium was studied by varying the LC type and the particle size, the velocity of the shearing of the KBr plates, and the layer thickness. As a quantitative indication about the degree of orientation of the sample and the limitation of the applying RDP for polarized IR-LD spectra interpretation, the parameter $\Delta$ can be used [34,35]. The quantitative approach to answering these questions lies in the calculation of the differences between intensive and negative bands in the corresponding differential IR-LD spectrum, or $[(A^{v1}_{\parallel} - A^{v1}_{\perp}) - (A^{v2}_{\parallel} - A^{v2}_{\perp})]$ for two absorption peaks, corresponding to integral absorbencies in the difference spectrum. The parameter $\Delta$ evidently depends on the parameter dichroic ratio. However, to obtain adequate conclusions both peaks must belong to nearly perpendicular transition moments, thus securing their different orientation toward the LC direction **n** and in the corresponding difference spectrum. As a model system we used DL-isoleucine, where the peaks at 1501 cm$^{-1}$ and 1417 cm$^{-1}$ are assigned to the NH$_3^+$-symetric bending ($\delta^s_{NH3+}$) and symmetric COO$^-$-stretching ($v^s_{COO-}$) modes (Figure 2.5). The maximal orientation of the suspended particles of all of the compounds studied was obtained at a cell thickness of 100 μm, shearing velocity of 3 μm/s, and particle size of 0.3 to 0.9 μm, using ZLI-1695 and MLC-6815 as hosts.

For experimental validation of the orientation technique, the classical scheme for analysis of liquid crystal colloids was used (Figure 2.6). As can be seen from the 3D graph in Figure 2.7, the maximal orientation is obtained in the particle size range 0.5 to 0.8 μm, at cell thickness of 100 μm, and shearing velocity of 3 μm/s for DL-isoleucine

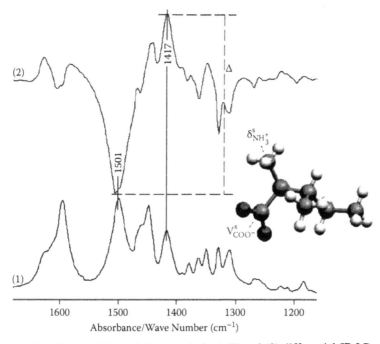

**FIGURE 2.5**   1700 to 1100 cm⁻¹ (1) nonpolarized IR and (2) differential IR-LD spectra of DL-isoleucine as an NLC suspension.

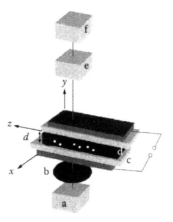

**FIGURE 2.6**   Experimental setup: (a) source, (b) polarizer, (c) KBr pellets, (d) nematic liquid crystal suspension, (e) analyzer, and (f) microscope.

the compounds studied. It is interesting to note that for DL-isoleucine a reasonable orientation is achieved for particle size of 1 μm, assuming that the orientation depends on the nature of the suspended particles, as has been proposed theoretically.

The role of the crystal class on the degree of orientation was examined by studying the three compounds with polymorph modifications. In all cases the monoclinic

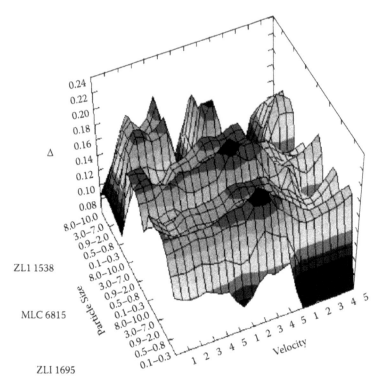

**FIGURE 2.7** 3D graph with the dependence of the parameter Δ versus particle size (μm), the velocity of the slippage (μm/s), and the layer thickness of the suspension of DL-isoleucine in the nematic liquid crystals ZLI-1695, MLC-6815, and ZLI-1538. **(See color insert.)**

and orthorhombic modifications are characterized by a higher Δ, but for the purpose of IR-LD spectroscopy, that is, IR-band assignment and structural information, the results obtained for all of the crystal classes are adequate for the purpose. For both polymorphs of 2-[5,5-dimethyl-3-[2-(2,4,6-trimethoxyphenyl) vinyl] cyclo-hex-2-enylidene] malononitrile, the Δ data show that the optimum degree of orientation is observed at a shearing velocity of 3 μm, cell thickness of 100 μm, and particle size within 0.3 and 0.5 μm. A reasonable orientation is achieved for aggregates with sizes between 1.0 and 8.0 μm. The triclinic form is characterized by a low degree of orientation of the suspended particles with a Δ difference of 0.30. Similar conclusions were also reached for the monoclinic and orthorhombic polymorphs of paracetamol. However, in these systems a reasonable orientation was observed at cell thicknesses of 100 μm. The tendency for a significant orientation of the samples was also retained for the α- and β-polymorphs of glycine, crystallizing with a monoclinic space group. From the IR-LD spectroscopic point of view, the trigonal (γ-) polymorph of glycine with space group $P3_1$ is oriented independently of the obtained difference of Δ = 0.08.

Like in all of the previously described systems, 5-amino-2-methoxypyridine ester amide of squaric acid ethyl ester with a monoclinic space group shows an orientation of the suspended particles under the above-mentioned experimental conditions. It is interesting, however, that in 5-amino-2-methoxypyridine ester amide of squaric acid ethyl ester, reasonable values of Δ are obtained at a cell thickness of 50 μm, with a Δ difference of 0.02. This means that at the lower concentration the suspended particles give a result capable of interpretation. Such a result is usually observed in layered systems, where the orientation of the suspended particles is along one of the crystallographic axes. In 5-amino-2-methoxypyridine ester amide of squaric acid ethyl ester only one intermolecular interaction of the NH...O=C type is formed with a bond length of 2.955 Å [35], and the observation in the difference IR-LD spectrum of a positive peak at about 1800 cm$^{-1}$ for symmetric stretching motion, $v^s_{C=O(Sq)}$ confirms the orientation of the layers along the $b$ axis (Figure 2.8).

A reasonable degree of macro orientation of the suspended particles of the tetragonal dipeptide L-alanyl-L-alanine is obtained in the particle size range of 0.5 to 0.8 μm at 100 μm cell thickness. It is interesting to note in this case that the sample orientation is sensitive to the velocity of the shearing of the KBr plates and the optimal conditions are within 3 ± 0.5 μm/s. As in the previous example, the hexagonal dipeptide L-phenylalanyl-L-phenylalanine yields maximum orientation at 3 μm/s, 100 μm cell thickness, and particle size within the range 0.1 to 0.9 μm, with a Δ difference between the particle size ranges between 0.02 and 0.03. An amorphous sample of the hydrochloride of homoglycyl-tetrapeptide is oriented at a significant level for a shearing velocity 3 ± 0.5 μm/s and particle size between 0.1 and 0.8 μm, and 100 μm cell thickness.

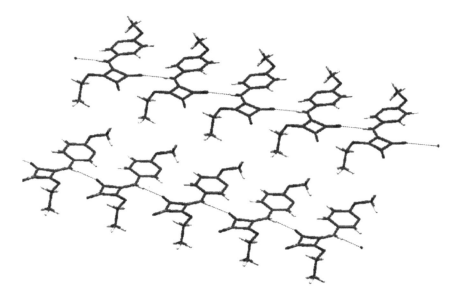

FIGURE 2.8 Hydrogen bonding in a crystal of 5-amino-2-methoxypyridine ester amide of squaric acid ethyl ester. (Koleva, B., T. Kolev, V. Simeonov, T. Spassov, and M. Spiteller. 2008, *J. Incl. Phen*, 61:319–333. With permission.)

A comparison of the degree of orientation of both polymorphs of 2-[5,5-dimethy l-3-[2-(2,4,6-trimethoxyphenyl)vinyl]cyclo-hex-2-enylidene] malononitrile oriented as colloidal suspensions in NLC and as polycrystalline samples oriented by melting between two KBr plates, a classical technique for orientation of solids, reveals that that about 15% to 20% of the particles are oriented in the CS-NLC method. These data correlate well with the electron microscopy observations (Figure 2.1) and can be explained by the fact that the sample contains a wide range of particle dimensions (100 nm to 10 μm) while the theoretical requirement for maximum orientation in the present study is for particles in the 0.5 to 0.9 μm range. The use of suspended particles of optimum size results in a degree of orientation of 88% to 91%, depending on the type of compound embedded. As can be seen in Figure 2.1 and Figure 2.7, in some cases we observed an orientation of agglomerates. Independent of this, the method is informative since the differential IR-LD spectrum only reveals an embedded compound that is oriented. In the case of nonoriented particles, the $A_{\parallel}$ and $A_{\perp}$ for a given $v_i$ are equal and subtracting both spectra at parameter 1 results in a zero-base line.

In all cases the prepared samples must be measured in the time interval of 0 to 120 min. After that, depending on the type of the suspended particles, a significant increase in the level of disordered particles is observed, accompanied by the formation of starlike structures for example, resulting in low values of $\Delta$.

An experimental full factorial design on two levels of variation of the input factors ($2^3$) was used to estimate the impact of the experimental parameters (size, shearing velocity, thickness) on the analytical signal for each system investigated. The information matrix of the design and the experimental results with assessment of the experimental bias (analytical signal, signal intensity) are discussed here. The three input factors $X_1$, $X_2$, and $X_3$ are, respectively, size, shearing velocity, and thickness. The two levels of variation of the factors normalized as +1 and −1 are as follows: $X_1$ (0.3 − 0.1 (−) and 0.9 − 2.0 (+)), $X_2$ (2 (−) and 4 (+)), and $X_3$ (50 (−) and 150 (+)), respectively.

The experimental design makes it possible to construct a polynomial regression model, which represents the impact of the single factors or their combination on the output function. Models for each of the systems considered were constructed and checked for variance homogeneity, regression coefficient significance, and model fit. The validity tests revealed that only a few regression coefficients were statistically significant (higher values than the significance number, which estimates the experimental error) and need to be taken into account in the data interpretation. The free term of the regression $a_0$ is a measure for the average signal intensity; the coefficients $a_1$, $a_2$, $a_3$ reflect the impact of the experimental parameters involved.

The maximal signal intensity was registered for the systems 2-[5,5-dimethyl-3-[2-(2,4,6-trimethoxyphenyl)vinyl]cyclo-hex-2-enylidene] malononitrile monoclinic form and L-alanyl-L-alanine (higher than 0.3), followed by 5-amino-2-methoxypyridine ester amide of squaric acid ethyl ester, orthorhombic and monoclinic forms of paracetamol, and polymorphs of the amino acid glycine systems (between 0.2 and 0.26). Lowest intensity was recorded for the systems dipeptide    L-phenylalanyl-L-phenylalanine, 2-[5,5-dimethyl-3-[2-(2,4,6-trimeth oxyphenyl) vinyl] cyclo-hex-2-enylidene] malononitrile triclinic polymorph and amorphous homoglycyl-tetrapeptide (lower than 0.2).

The experimental factor "size" shows significance for only three systems (L-phenylalanyl-L-phenylalanine, 5-amino-2-methoxypyridine ester amide of squaric acid ethyl ester, and γ-polymorph of glycine) and its impact is always negative. It may be concluded that its role in the signal formation is not substantial except for the system 5-amino-2-methoxypyridine ester amide of squaric acid ethyl ester where lower size causes a better output signal.

Four out of eleven insignificant regression coefficients—for (2-[5,5-dimethyl-3 -[2-(2,4,6-trimethoxyphenyl)vinyl]cyclo-hex-2-enylidene] malononitrile triclinic polymorph, the orthorhombic and monoclinic forms of paracetamol, and the hydrochloride of the homoglycyl-tetrapeptide)—are marked for the second experimental factor (shearing velocity). The effect of $a_2$ is predominantly negative, which means that a lower shearing velocity offers opportunity for a higher signal.

Only the system L-alanyl-L-alanine indicates insignificance with respect to the impact of the input factor "layer thickness" on the output signal. Therefore, this is the most important experimental factor for almost all systems. Its effect is predominantly negative, which means that better signals are obtained for a lower layer thickness.

In this study we face a problem, which can also be handled by the use of multivariate statistical approaches. Each of the systems involved was successfully described by a row of reliable descriptors, for example, the regression coefficients from the models representing the impact of each experimental parameter on the output signal. When cluster analysis was applied to the results (11 objects, each one of them described by four coefficients having their own physical meaning and impact) one should find levels of similarity (or dissimilarity) between the systems studied. Cluster analysis is a very important classification and projection method, which makes it possible to plot a multidimensional system on a plane and to detect groups of similarity (clusters) on the plot. In our case, we applied hierarchical agglomerative clustering of the normalized input data (z-transformation), squared Euclidean distance as a measure of similarity, and Ward's method of linkage between the variables.

A high level of similarity is observed between all polymorphs of glycine and paracetamol. The L-alanyl-L-alanine object could also be included within the big cluster glycine and paracetamol. This means that all these experimental systems behave in quite a similar manner with respect to their analytical signal and experimental parameter impacts. They are characterized by a medium level of relative signal intensity and do not exhibit a very strong impact of the experimental parameters. In a way these systems could be easily exchanged with one another in reaching a certain signal intensity despite the various experimental conditions. The salt of the homoglycyl-tetrapeptide system also belongs to this cluster.

The systems L-phenylalanyl-L-phenylalanine and 2-[5,5-dimethyl-3-[2-(2,4,6-trimethoxyphenyl)vinyl]cyclo-hex-2-enylidene] malononitrile monoclinic also create a group of similarity because of their very strong dependence on the shearing velocity parameter. Although having different relative intensity values, they prove to be similar on the basis of a strong $a_2$ impact (positive influence). The relatively high (although negative) influence of the thickness parameter also plays a role in this case.

No other system is influenced so significantly by the simultaneous impacts of the layer thickness and shearing velocity parameters.

The systems 5-amino-2-methoxypyridine ester amide of squaric acid ethyl ester and 2-[5,5-dimethyl-3-[2-(2,4,6-trimethoxyphenyl)vinyl]cyclo-hex-2-enylidene] malononitrile triclinic polymorph differ from the other two groups of similarity and occupy the position of outliers rather than belonging to a separate group. They indicate quite different intensities and impacts of the input parameters, and should be considered as individual cases with respect to their output signal and response to the experimental conditions.

Since a clustering of the systems in consideration was obtained, it was important to check if the system descriptors (regression coefficients) are correlated in one way or another. The cluster analysis with respect to the variables yielded the dendrogram.

Two distinct groups of descriptors are formed. It may be assumed that two major factors are responsible for the linkage between the system descriptors: the first one indicating linkage between relative intensity and velocity (conditional name "dynamic" experimental factor) and the second one (conditional name "static" experimental factor) demonstrating the correlation between size and thickness parameters. Both groups are quite different in their impact on the system's response.

## 2.2.1 Accuracy and Repeatability

The repeatability was determined by $v_i$ and $A_i$ at 1417 cm$^{-1}$ and 1501 cm$^{-1}$ peaks of 10 parallel independent samples measured at 150 scans, Lorentzian to Gaussian peak function ratio at 50:50, $\chi^2$ factors between $0.0004_2$ and $0.0004_1$, and 2000 iterations. Three repeated measurements were obtained for each sample as well. The peak positions at 1501 cm$^{-1}$ $\pm 0.02_2$ and $1416.97 \pm 0.02_4$ cm$^{-1}$ were obtained. The reproducibility of the integral absorptions was determined by $A_{1501}/A_{1417}$ ratios, where the results indicated a $0.452 \pm 0.032_8$ value.

## 2.2.2 The Quantitative Ratio of Liquid Crystals and Solid Samples

The viscous characters of the suspension and the liquid crystal medium at room temperature (25°C) require using a weight percent unit for quantitative analysis. The determination of the weight percent interval is important because, depending on the liquid crystal quantity, the orientation order ($S$) varies within the limits 0 to 1. Furthermore, a lower degree of orientation of the sample leads to the impossibility IR-LD spectral analysis and application of RDP. The quantitative approach to answering the posed question is to calculate $[(A^{v1}_{\parallel} - A^{v1}_{\perp}) - (A^{v2}_{\parallel} - A^{v2}_{\perp})]$ for two absorption peaks, corresponding to integral absorbencies in the difference spectrum. However, to obtain adequate conclusions both peaks must belong to near to perpendicular transition moments, thus securing their different orientation toward the liquid crystal direction **n** and in the corresponding difference spectrum. In our model system, the peaks at 1501 cm$^{-1}$ and 1417 cm$^{-1}$ assigned to the NH$_3$$^+$-symmetric bending ($\delta^s_{NH_3^+}$) and symmetric COO$^-$-stretching ($v^s_{COO^-}$) modes [34,35] are used for this purpose.

The transition moments of $\delta^s_{NH3+}$ and $v^s_{COO-}$ are mutually oriented, at an angle at 109.3(7)° and causing the observed negative and positive orientation of the corresponding peaks in the difference spectrum [34]. A significant opposite orientation of both discussed peaks is observed at 5 ± 1 weight percent, where the absolute values of the differences $(A^{v1}_{\parallel} - A^{v1}_{\perp})$ and $(A^{v2}_{\parallel} - A^{v2}_{\perp})$, respectively, are remarkable. The obtained results correlated well with one another, indicating an orientation parameter $S \rightarrow 0$ in the cases of the largest and lowest amount of the guest molecule. These results were obtained by immediate scanning, avoiding the possibilities of any additional effects and especially the orientation quality changes caused by delaying the measurements.

### 2.2.3 PRELIMINARY RUBBING OUT OF THE KBR PLATES

The rubbing out in one direction of the used KBr plates with sandpaper, provides a supported orientation effect of the guest compound. The discussed additional effect was examined using the integral absorbencies of the $v_{CN}$ stretching peak at 2235 cm$^{-1}$ in parallel and perpendicular IR-LD spectra of pure liquid crystal ZLI-1695 varying as well as the type of sandpaper. The data indicated a significant degree of orientation in the case of KBr plates rubbed out in advance with sandpaper size C800. In the other cases, unsatisfactory effects were established, which may be explained with the molecule field influence of the liquid crystal medium, resulting in a minimal orientation effect both without sandpaper and in the case of rubbing out with coarse sandpaper.

### 2.2.4 PEAK FUNCTION TYPE FOR THE CURVE-FITTING PROCEDURE

The determination of $v_i$ and $A_i$ by curve fitting requires three possible types of peak functions (Lorentzian, Gaussian or mixed Gaussian, and Lorentzian peak function). To determine the best one, our method for orientation obtains $v_i$ and $A_i$ of the peak at 1417 cm$^{-1}$ in DL-isoleucine measured as an NLC suspension. Dependencies of $\chi^2$ versus Lorentzian to Gaussian peak function ratios in percentages are presented. The minimal $\chi^2$ value of $0.0007_3$ is obtained at a 50:50% Lorentzian:Gaussian peak function ratio.

### 2.2.5 NUMBER OF SCANS IN THE MEASUREMENTS

To determine the number of scans in the CS-NLC measurements, $A_{max}$ values of series of peaks versus the number of scans are obtained and a "saturation" in all the cases is established after 140 to 150 scans. Therefore, the standard number of scans in all IR-LD measurements using the presented method is 150.

## 2.3 EXPERIMENTAL DESIGN

The experimental design is made using the following different input experimental factors (experimental conditions to maintain the positive and negative levels): amount of the compound studied ($X_1$); the degree of rubbing out of KBr pellets ($X_2$);

the number of scans ($X_3$); and the ratio of Gaussian and Lorentzian peak functions in the curve-fitting procedure ($X_4$) on the spectroscopic signal for five different frequencies—2235 cm$^{-1}$, 1512 cm$^{-1}$, 1501 cm$^{-1}$, 1493 cm$^{-1}$, and 1413 cm$^{-1}$.

- 2235 cm$^{-1}$ peak (Strong Overlapped Frequency)—It is readily seen that all coefficients for the first signal are statistically nonsignificant. This means that no effect of influence of the input factors $X_i$ could be found (the experimental error of repeatability is, in deed, higher that the calculated regression coefficients).

  For the second signal of the same matrix frequency, four coefficients indicate high significance, namely, those related to $X_1$ (amount of chemical compound) and $X_2$ (number of scratches). Some of the mixed coefficients are also significant, like $a_{23}$ and $a_{34}$, which is an indication of a nonlinear relationship, and simultaneous effects of the parameters number of scratches and number of scans or number of scans and parameter $X_4$. In principle, the mixed interactions affect the signal relatively weakly. Nevertheless, the significance of the mixed effect $X_3X_4$ indicates a synergetic effect of time and number of scans being higher than the separate influence of the single parameters.

- 1512 cm$^{-1}$ peak (Strong Overlapped Frequency)—In this situation many regression coefficients, even for the first signal, are statistically significant. The important role of the input parameters 2, 3, and 4 is stressed (only parameter 1 is close to nonsignificance). Parameter 3 is characterized by its negative effect to the signal (for both signals 1 and 2). It means that the number of scans should be reduced in order to enhance the signal (the same effect is generally observed for the matrix frequency 1 but for nonsignificant coefficients).

  Quite a large number of mixed coefficients are significant. This is an indication for the nonlinear regression relation between the signal and the experimental parameters. The special role of parameter 2 is also confirmed for this frequency; it has the highest positive value and, obviously, affects most seriously the signal registered. Also, mixed coefficients with participation of parameter 2 show high significance: $a_{12}$, $a_{24}$, $a_{124}$, $a_{234}$. Parameter 4 is also significant in this case.

- 1501 cm$^{-1}$ peak (Overlapping Frequency)—The behavior of this system is quite similar to the previous one. There are, however, some substantial differences. For instance, $a_4$ is negative, which means that for this particular frequency the parameter $X_4$ has an opposite effect as compared to frequency 2. The mixed effects are almost the same as in the previous system, which is, again, an indication of nonlinear interactions and synergetic effects of the single input parameters.

- 1493 cm$^{-1}$ (Overlapping Frequency)—Again, a series of differences is observed. The specific point is that $a_0$, which reflects the average signal, has a higher level for signal 1 as compared to signal 2 (at all other frequencies the situation was reversed). More coefficients are significant for signal 1 than for signal 2, which is also different than the other cases. This "turnover" situation probably corresponds to the character of the spectrum. Otherwise, the significance of the input parameters "guest molecule quantity," "number of scratches," and "time" (the last one, again, with negative impact on the output signal) is shown.

- 1413 cm$^{-1}$ (Nonoverlapping Frequency)—This case is quite different as compared to the other frequencies. It resembles to a certain point the case with frequency 1 as the coefficients for signal 1 are all insignificant. A specific aspect is also the negative sign of $a_1$ and $a_2$ (both significant for signal 2), which is completely different as compared to all other frequencies. This is the only case where the coefficient $a_{1234}$ possesses significance, giving an indication of a very high order of interactions.

The experimental design carried out made it possible to determine in a qualitative way the impact of four input experimental factors (experimental conditions) on the spectroscopic signal for five frequencies. All the data obtained correlated with the validation discussed earlier. The following general conclusions can be offered:

- The *material quantity* factor is very significant for all frequencies in consideration (except for the first signals in frequencies 1, 2, and 5). When statistically significant it shows a positive impact on the output signal (mainly signal 2 for the various frequencies). The only exception is the negative sign of $a_1$ for the specific frequency 5 where the "material quantity" factor increase worsens the signal.
- The *preliminary rubbing-out* factor is significant for most of the systems with positive impact on the output signal. Again, the only exception is frequency 5 where the impact has a negative sign.
- The *number of scans* factor does not indicate constant or serious impact on the output signal except for frequency 2, where it contributes to the diminishing of the signal with the increasing number of scans.
- The *Lorentzian and Gaussian peak functions ratio* factor (with the exception of frequencies 1 and 5) is always of serious importance, having a positive impact for frequency 2, and negative impact for frequencies 3 and 4. It seems that this particular factor reveals no specific impact but rather a random one.
- The mixed interactions are not highly significant but there are some indications for their serious impact, especially due to the simultaneous and probably synergetic effects of all inputs.
- In order to achieve an optimization of the whole system under consideration, one has to select very carefully and successively the conditions for the sample preparation (material quantity, number of scratches) and the signal reading (number of scans, time, frequencies).

Validation of RDP based on subtracting the nonpolarized IR spectra of *n*-component solid mixtures was presented. The accuracy and precision were established. The limits of detection are 3.0, 2.5, 1.5, and 1.0 wt.% for five-, four-, three-, and two-component mixtures, respectively. Smoothing procedures for IR spectral analysis, based on Savitzky–Golay or Fourier methods, were applied as well. The mean values and relative standard deviations for peak position ($v_i$) and integral absorbance ($A_i$) obtained by this data processing approach have been examined using Student's *t*-test.

## 2.4 MATHEMATICAL TOOLS FOR THE INTERPRETATION OF THE IR SPECTROSCOPIC PATTERNS

### 2.4.1 IR Spectra Subtraction

For an *n*-component system [33], the subtraction [73,74,121] of the IR spectra is obtained from the relation

$$RS = SS - (SbSF) \tag{2.5}$$

In Equation 2.5, RS is the *result spectrum*, SS the *sample spectrum*, SbS the *subtracted spectrum*, and *F* the *subtraction factor*. By varying the *F* factor and looking at the obtained *reducing curve*, the disappearance of a given absorption peak can be observed. However, this procedure entails the problem concerning shifting the data point spacing sets. The proper subtraction of one spectrum from another requires a strict correspondence of the data points to the same *x* values. The *data point spacing* problem needs interpolation, which is done automatically by software. The software also handles the problem of the subjectivity of iterative algorithm selection. This algorithm calculates the subtraction factor by minimizing the complexity of the residual spectrum (*auto subtraction algorithm*).

### 2.4.2 The Smoothing Procedure

The most typical smoothing methods for IR spectral curves are Savitzky–Golay [122–125] and Fourier [122,123]. Fourier smoothing is based on data inversion by *fast Fourier transform* (FFT) to the time domain, where a trapezoidal filter is applied to the high frequency region. Finally, an inverse FFT is applied to give the smoothed result. The degree of smoothing determines the cutoff point of the filter. Savitzky–Golay smoothing is based on the convolution approach, which performs a least squares fit to a specified part of data points. The degree of polynomial and the number of points are controlled during the procedure. The algorithm used for generating the Savitzky–Golay convolution coefficients has been described [123–125]. It is worth mentioning that the choice of smoothing parameters is somewhat subjective—the data are dependent and arbitrary.

#### 2.4.2.1 Limits of Detection and Smoothing Procedure Validation

The limits of detection (LODs) of the RDP for nonpolarized IR spectra have been examined for *n*-component systems starting with optimization of the standard smoothing conditions for IR curve interpretation. For this purpose the complicated IR spectrum of a five-component mixture (for example, glycyl-L-methionyl-glycine, L-tyrosine, L-phenylalanine, L-phenylalanyl-L-phenylalanine, and DL-isoleucine) was stepwise reduced by subtracting the IR spectra of other components. In this case the *reduced IR spectrum* should correspond to pure DL-isoleucine. A series of IR characteristic peaks are used in the following steps:

1. The IR spectrum of sample 1 is subtracted from the IR curve of pure glycyl-L-methionyl-glycine until disappearance of the peak at 3295 cm$^{-1}$ to yield a final reduced IR spectrum.
2. The L-tyrosine IR spectrum is subtracted from the latter curve at the maximum at 3205 cm$^{-1}$.
3. A subsequent subtraction, where the subtracted spectrum corresponds to L-phenylalanine, results in an IR curve of the reduced IR spectrum. In this case the subtraction is aimed at disappearance of the 1624 cm$^{-1}$ peak, typical for L-phenylalanine. However, the complex character of the spectral curve calls for monitoring of a second peak, also typical for a simple amino acid. For L-phenylalanine, the 844 cm$^{-1}$ band is suitable.
4. Finally, the last spectrum is subtracted from the IR curve for individual L-phenylalanyl-L-phenylalanine until total elimination of the peak at 1495 cm$^{-1}$, which corresponds to pure DL-isoleucine in solid state, is reached. The comparison between the first and last curve demonstrates a qualitative correlation. It is noteworthy that the order of subtraction of different components in the mixture leads to identical final results.

However, the exact qualitative or quantitative IR spectra identifications of unknown compounds in mixtures are difficult, since the path lengths in all solid-state techniques are uncertain. For that reason, the analysis requires a comparison between the pairs of IR spectral curves (standard, pure compound, and reduced IR spectra) in regard to the peak position and integral absorbance ratios of a series of IR characteristic peaks. In DL-isoleucine the multiple absorption maxima in the 1700 cm$^{-1}$ to 1500 cm$^{-1}$ region were chosen, where the deconvolution and subsequent curve fitting yield three absorption peaks. The reduced (components $n = 5$) and pure IR spectrum of the compound were interpreted using the Fourier and Savitzky–Golay smoothing procedures with variable smoothing degrees between 0 to 10 (with step 1) and 5 to 25 (with step 2) in the first and second methods, respectively.

The results for $v_i$ and $A_i$ were examined using the Student's $t$-test. The $p$ values varied between 0.17 and 0.32 using Fourier smoothing (degrees 1 to 5) and were equal to 0.36, when the IR spectral curve was smoothed by the Savitzky–Golay method (degree 5). The calculated $p$- and $t$-data are compared with tabulated data [68–70] at a 99% confidence level to verify the absence of statistically significant differences [70]. This is seen when the standard deviations (SDs) of unprocessed and Fourier-smoothed (degree 5) IR spectra of DL-isoleucine are presented. These results are well correlated with other known data from quantitative IR analysis [126,127]. In other cases of higher degrees of smoothing (7 to 25 by Savitzky–Golay and 6 to 9 by Fourier), critical $p$- and $t$-values were obtained, thus leading to rejection of the hypothesis about the possibility of adequate comparison of the resulting curves [126,127]. For example, the comparison between the original, Savitzky–Golay (degrees 7, 9, and 25), and Fourier (degree 9) smoothed IR spectral curves of DL-isoleucine gives $p$

values around the critical value of 0.05. Despite the slightly higher $p$ values of 0.076 and 0.072 for degrees 7 and 9 by Savitzky–Golay, these higher degrees of smoothing cannot be recommended for quantitative assays. The inadequate SDs for original and Fourier smoothed data (degree 9) also tend to support cautiousness.

These results lead to the following main conclusions: (1) The best correlation between original and smoothed IR spectral data of DL-isoleucine was obtained using the Fourier method with degrees of smoothing between 1 and 5 and by the Savitzky–Golay procedure only with degree 5. (2) As far as optimization of smoothing conditions is concerned, adequate results were only obtained for a five-component mixture for components of at least 3 wt.%. This value should be accepted as the limit of detection for the discussed system. The second conclusion is supported by the following control study. Examination of the second 3 wt.% component glycyl-L-methionyl-glycine versus a series of absorption peaks, determined also by deconvolution and the curve-fitting procedure, was carried out. A comparison with standard glycyl-L-methionyl-glycine IR spectral data was then made. The calculated integral absorbance ratios and $t$-test data allow conclusions similar to those for DL-isoleucine—likewise the SDs for $n = 5$.

### 2.4.3  Accuracy and Precision

The accuracy and precision were obtained for 17 $n$-component systems in the cases of the best (lowest) LODs, namely, samples 1, 6, 10, 14 corresponding to $n = 5$-, 4-, 3-, 2-component systems and for standards 16 and 17 for $n = 1$. All the spectral curves are Fourier smoothed at degree 1. The data obtained included the calculated SDs. The SDs for $v_i$ are obviously lower than those for their corresponding values for $A_i$. The result is in agreement with Thulstrup and Eggers [5], and could be explained by uncertainties involved by the curve-fitting stage for complicated absorption maxima. The SDs for $v_i$ and $A_i$ for DL-isoleucine are: $v_1 = 1627.80 \pm 0.03_4$, $v_2 = 1613.69 \pm 0.01_6$, $v_3 = 1593.60 \pm 0.02_0$, $A_1/A_2 = 9.436_1 \pm 0.011_4$, $A_1/A_3 = 0.450_3 \pm 0.015_9$ and $A_2/A_3 = 0.046_5 \pm 0.015_8$, respectively.

### 2.4.4  Deconvolution

As a mathematical tool, deconvolution is used for the interpretation of strongly overlapped IR spectroscopic bands. The method allows a definition of the number of submaxima of a given multicomponent IR band. The deconvolution uses the following methods: Fourier self-deconvolution (FSD) and noise filters. Fourier self-deconvolution is a special high pass FFT filter that synthetically narrows the effective trace bandwidth features. This aids in identifying the principal bands that make up a more complex band with overlapping features. This can be useful for more accurate determination of the number of peaks in a trace region, the band positions, and areas. This technique can also be used to accurately determine starting parameters for applications such as curve fitting.

This application is based on the method described by Griffiths and Pariente [446,447]. Two filters are employed. An exponential filter is used to sharpen spectral features; the constant $\gamma$ is varied to change the filter shape. The $\gamma$ parameter equals the full width at half height (FWHH), (or where X coordinates are found by interpolating between the points just before and after the half height points) of the widest resolvable peak. It should be realized that the imposed filter is the transform of a Lorentzian line shape.

As FSD tends to increase the apparent noise in the data, there is some benefit to be gained by simultaneously applying a low-pass smoothing filter. This can effectively reduce the noise, which may be at a higher spatial frequency than the peaks, without losing peak resolution. The forms of the filters are boxcar and Bessel, and are mathematically described in the following.

The deconvolution filter is a simple exponential filter of the form $e^{2\pi\gamma X}$ where $\gamma$ is the deconvolution filter constant and X is the array (i.e., data file) whose X-axis range is normalized between 0 and 1. This function is multiplied by the Fourier transformed trace, and the data is then reverse Fourier transformed to give the result.

The noise filter procedure consists of selecting the appropriate filter parameters, which is critical in obtaining good (and valid) results with FSD.

The Bessel smoothing function ($x$) takes the form

$$[1 - (x/X)^2]^2 \qquad\qquad (2.6)$$

where $0 \leq x \leq X$, and X is an array (i.e., data file) where the X-axis range is normalized between 0 and 1. Thus, the Bessel filter defines the fraction of the Fourier transformed data to be zeroed and provides a smooth rolloff. The boxcar smoothing function simply zeros all points outside the cutoff.

Figure 2.9 illustrates the IR spectroscopic pattern with and without the applied FSD procedure. The data shows well-defined pairs of submaxima of the broad absorption band within the IR spectroscopic region of 1475 to 1440 $cm^{-1}$.

### 2.4.5 CURVE-FITTING PROCEDURE (CFP)

There are many ways to write algorithms to detect peaks in data, and no one method is perfect. In fact, the best peak picker in the world is actually the human eye. However, since scientists often rely on computers to perform basic analyses of data. Thermo Galactic GRAMS® AI/7.01 (Thermo Galactic, United States), an IR spectroscopic software package, has developed its own peak-picking algorithms. The algorithms and techniques described next are general descriptions of the peak-picking methods used in all Thermo Galactic software. Note that this merely describes a straightforward peak locating and calculation method. It does not work for poorly resolved peaks, which usually require more sophisticated peak determination methods such as FSD or nonlinear peak fitting (NPF).

The algorithm is essentially a multipass data scanning method: Scan the data array and calculate the delta Y between adjacent pairs of points for the entire spectrum and take the average. This is considered the "noise value." Scan the data array again, then do the following: Look for data points where the delta Y is greater than the noise value and the delta Y value is positive. This is the start of a peak. Continue scanning and look for a local maximum point (Y values of the data points

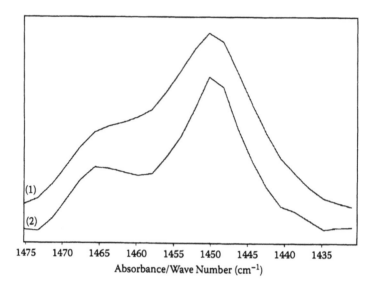

**FIGURE 2.9** (1) Nonprocedured and (2) deconvoluted IR spectra of the liquid crystal MLC-6815 within the 1475 and 1430 cm$^{-1}$ IR region ($\gamma = 3.8$ and increment 0.1). The procedure is performed by GRAMS AI/7.01, an IR spectroscopic software package.

immediately to the left and right are both less than the point). This is considered the peak top. Calculate the "true" peak position using a center of gravity calculation on the top 3 points. Continue scanning until the absolute value of the delta Y is less than the noise value. This is the end of the peak. Continue scanning from this point until the end of the data is reached. Apply the edge detection parameters to refine the peak edges and baselines (Smoothing, Edge Detection, Minimum Edge Slope, Maximum Peak Group Separation, Maximum Peak Shoulder Ratio). Rank the peaks in order from biggest to smallest in terms of height and apply peak rejection parameters (Sensitivity, Threshold, Minimum Area) to discard unwanted peaks.

One of the more difficult problems in science is finding an equation that fits the shape of a curve that is represented by more than one functional shape or form. Since there are many solutions to fitting even two equations to one curve, "plug and chug" or linear regressions will not work. The solution needs to be the best combination of all the parameters for the two (or more) functions. In general, nonlinear peak fitting methods involve fitting a series of individual functions simultaneously to obtain the single best fit solution. How this best fit (merit criterion) is calculated and how the fitting variables are adjusted are the basic differences between the methods.

The solutions are found by iteratively trying a series of combinations of the parameters until the best one is found (as predetermined by the merit criterion). Since the solutions are interdependent, small changes in one of the parameters affect the final result of all the others. When using an iterative process, the starting point should be as close to the actual solution as possible. Good "guesstimates" for the starting values increase the probability of finding the best solution.

As always, no method is perfect and, unfortunately, since there are a number of answers to a nonlinear problem, on many occasions the fit will end up in a "local minimum," which may not be the best possible solution. Many fit routines will then continue to iterate using solutions that are significantly different from the minimum, trying to find another better solution. If one is not found, then the minimized solution is considered best.

Other problems occur if the user input values are significantly far from the "real" answer. This may end in a solution that is stuck in a local minimum that is determined to be the best fit. In extreme cases the fit may even "walk away" to a ridiculous solution. The only way to recover from this problem is to reenter the starting fit values. Overfitting is another common problem. Better fits can always be obtained with more input variables (given enough variables you can fit anything). Thus it is important that the input parameters reflect the proper number of variables based on the physical measurement.

Using nonlinear methods requires a threshold at which the fit is considered good (i.e., minimizing the merit equation to some near-zero value). In the case of the Levenberg–Marquardt method, the merit equation used is $X^2$. The final solution is found when a minimum in the reduced $X^2$ equation is reached. It is a statistical measure of goodness-of-fit, inversely proportional to the known variance of the data set. The published tables give the threshold for various degrees of freedom and different significance levels. The tables, however, are based upon the assumption that the error or uncertainty is known exactly, which is usually not the case.

Figure 2.10 shows the curve-fitted IR spectroscopic patterns of the nematic host MLC-6815 within the 1395 to 1355 cm$^{-1}$ region. The data shows two absorption

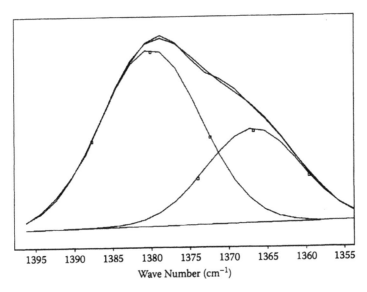

**FIGURE 2.10**   CF IR spectra of the liquid crystal MLC-6815 within 1395 to 1355 cm$^{-1}$ IR region (Lorentzian and Gaussian peak functions ratio 50:50 and 2000 interactions). The procedure is performed by the GRAMS AI/7.01 software.

subbands similar to the data obtained by the deconvolution procedure. The procedure allows defining of the integral absorbances and the peak position of each of the elucidated IR bands.

## 2.4.6 BASELINE CORRECTION (BLC)

The BLC performs several different types of baseline leveling operations on a spectrum: two-point level, multipoint level, function fit, interactive polynomial, or auto level. These five methods are capable of baseline correcting a wide variety of data. Note that in the two-point level and the multipoint level methods, the baseline is leveled at a value that is the average of the baseline points. This allows the methods to be applied to both transmission and absorbance spectra. For the interpretation of both nonpolarized and polarized IR spectra, the two-point and the multipoint methods are usually used.

Plot a connecting line through the two selected points, as shown in Figure 2.11, then subtract it from the trace. The baseline is calculated using the standard linear equation:

$$y = Mx + B \tag{2.7}$$

where

$$M = \frac{y_2 - y_1}{x_2 - x_1}$$

$$B = y_1 - M.x_1 \tag{2.8}$$

For the multipoint baseline correction method, baseline level the connecting lines through the selected points and subtract them from the trace, as shown in Figure 2.12.

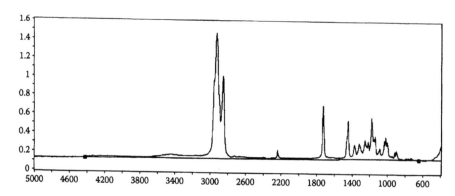

**FIGURE 2.11** Two-point BLC of the IR spectrum of liquid crystal MLC-6815. The procedure is performed by the GRAMS AI/7.01 software.

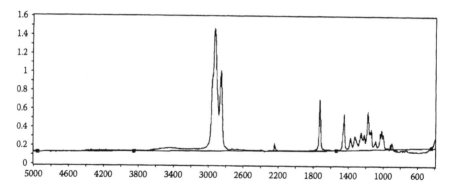

**FIGURE 2.12**   Multipoint BLC of the IR spectrum of liquid crystal MLC-6815. The procedure is performed by GRAMS AI/7.01 software.

## 2.5   REDUCING-DIFFERENCE PROCEDURE (RDP) FOR IR-LD SPECTRA INTERPRETATION

The interpretation of the nonpolarized and polarized infrared spectra includes determination of the position $(v_i)$ and integral absorbancy $(A_i)$ for each $i$-peak by FSD and CFP at a 50:50 ratio of Lorentzian to Gaussian peak functions. Usually, the $\chi^2$ factor varies between 0.00013 and 0.00008 (in the CS-NLC case $1.2.10^{-5}$ to $2.3.10^{-4}$) and 2000 iterations [76]. The means of two treatments were compared by the Student's $t$-test.

Spectroscopic and structural results of the presented orientation technique were obtained using the known RDP designated as "stepwise reduction" for polarized IR spectra interpretation. This method was initially suggested by Thulstrup and Eggers [5] for the interpretation of polarized UV spectra. The procedure involves consecutive elimination of the spectral bands of a given polarization by subtracting the perpendicular spectrum multiplied by a coefficient from the parallel one. This procedure was extended by Spanget-Larsen [62] and by Korte and Schrader [67] to samples orientated in stretched polyethylene and in nematic solution, respectively. A systematic analysis of this approach and its application to IR band assignment according to their symmetry appurtenance was developed by Jordanov [14–24,66] for polarized IR-LD spectra in NLC solution. The method consists of subtraction of the perpendicular spectrum ($IR_\perp$; resulting from a 90° angle between the polarized light beam electric vector and the orientation of the sample) from the parallel one ($IR_\parallel$) obtained with a colinear mutual orientation. The recorded difference ($IR_\parallel - IR_\perp$) spectrum divides the corresponding parallel ($A_\parallel$) and perpendicular ($A_\perp$) integrated absorbencies of each band into positive values originating from transition moments, which form average angles with the orientation direction **n** between 0° and 54.7° (magic angle), and negative ones corresponding to transition moments between 54.7° and 90°. In the RDP, the perpendicular spectrum multiplied by the parameter $c$, is subtracted from the parallel one and $c$ is varied until at least one band or set of bands is eliminated. The simultaneous disappearance of these bands in the reduced IR-LD spectrum ($IR_\parallel - cIR_s$) indicates colinearity of the corresponding transition moments, thus yielding information regarding the mutual disposition of the molecular fragments.

The presented determinations are the results of the following equations, definite for *i*-band with $v^i$ ($i = 1...n$):

$$(A^i_p - A^i_s) = A^{iso}.S.(3/2\cos^2\theta - \tfrac{1}{2})$$   (2.6)

where $S$ is the orientation factor, showing the average orientation of the molecule and varies between 0 and 1. If the $S = 0$, the effective orientation of the molecule is not observed or if $S = 1$ there is a maximal orientation. It is known that in the technique of orientation as an NLC solution, $S$ depends on the geometry and size of the guest molecule, temperature, and especially on the amount of the mesomorphic medium; $A^{iso}$ is the integral absorption of the *i*-band in the corresponding nonpolarized spectrum; and $\theta$ is the angle between the *i*-transition moment and the orientation director **n**. If $\theta = 0$, then

$$(A^i_p - A^i_s) = A^i, (A^i_p - A^i_s) > 0$$   (2.7)

and the corresponding peak is positive. For $\theta = 90°$, the

$$(A^i_p - A^i_s) = A^i S, \text{ and } (A^i_p - A^i_s) = -1/2\, A^i S < 0$$   (2.8)

leads to a negative oriented peak. If $\theta = 54.7°$, then

$$A^i_p = A^i_s$$   (2.9)

and in the difference IR-LD spectrum the peak is not observed. The last dependencies and conclusions may be illustrated using the polarized IR-LD spectra of the nematic host. The positive peak for the difference spectrum maximum at 2235 cm$^{-1}$ ($v_{CN}$) arises because of the $v_{CN}$ transition moment making an angle between 0° and 54.7° with **n**. In contrast the negative peak at 1450 cm$^{-1}$ corresponds to $\delta^s_{CH2}$, whose transition moment makes an angle between 54.7° and 90° with **n** (see Equations 2.8 and 2.9).

The interpretation of IR spectra containing multiple or strong overlapped peaks not only in polarized but also as in conventional spectra requires as a first step, the determination of the number of peaks; and as a second step, curve fitting resulting in $v_i$ and integral absorption $A_i$ values [36–39]. The deconvolution based on the FSD method and the subsequent peak-fitting procedure is applied in this respect (see Section 2.4.4).

The polarized measurements are made in the classical manner [3,4,6,7]. Hence, the polarized IR-LD spectra recorded are the perpendicular spectrum, ($IR_\perp$; resulting from a 90° angle between the polarized light beam electric vector and the orientation of the sample), and the parallel spectrum ($IR_\parallel$) obtained with a colinear mutual orientation together with a nonpolarized spectrum of the partially oriented suspension. In some cases the oriented samples can exhibit birefringence effects, which are difficult to eliminate. They can be minimized by placing the sample before the polarizer [3,4].

The application of the RDP for polarized IR-LD spectra interpretation is illustrated by the analysis of the IR-LD spectra of the nematic liquid crystal NP-1083® (Merck Darmstadt, Germany) depicted in Figure 2.13.

The corresponding difference IR-LD spectrum, depicted in Figure 2.14, is characterized by the positive IR band at 2229 cm⁻¹, belonging to the stretching vibration of the C≡N group ($v_{CN}$). In parallel vibrational modes, the bands at 1605 cm⁻¹ and 1502 cm⁻¹ (Figure 2.14) are also positive. These bands correspond to in-plane **8a** and **19a** modes of the substituted benzene ring in the molecule of the NLC and possess $A_1$ symmetry class (in the terms of the local $C_{2v}$ symmetry). The

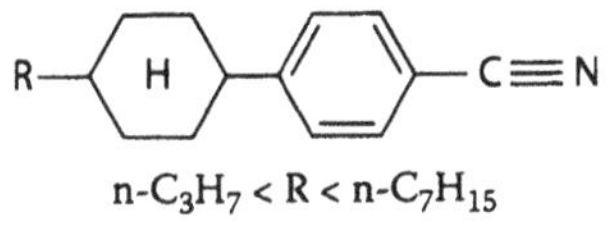

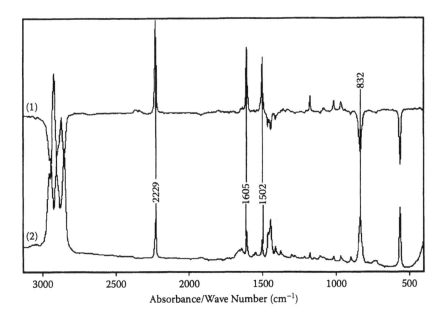

**FIGURE 2.13**  Chemical diagram of the NLC NP-1083. (Courtesy of Merck Darmstadt, Germany.)

**FIGURE 2.14**  (1) Differential IR-LD and (2) nonpolarized IR spectrum of NLC NP-1083. (Courtesy of Merck Darmstadt, Germany.)

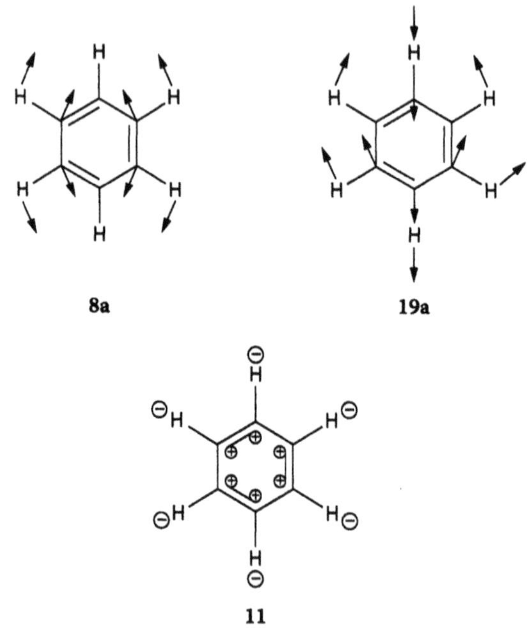

**FIGURE 2.15**   Normal vibrations **8a**, **19a**, and **11** of the benzene molecule.

resulting transition moments of these vibrations coincide with the main symmetry axis of the molecule (Figure 2.15).

That is why the IR bands at 1605 cm⁻¹ and 1502 cm⁻¹ possess an equal sign in the difference IR-LD spectrum and will be eliminated at equal DR in the corresponding reduced IR-LD spectrum (Figure 2.16). The IR band at 2229 cm⁻¹ is also eliminated. The direction of the transition moment of the stretching vibration of the C≡N group ($v_{CN}$) (Figure 2.13), also coincides with the main symmetry axis of the molecule, thus determining the elimination of the corresponding bands with IR maxima belonging to the **8a** and **19a** modes. The IR band at 832 cm⁻¹, corresponding to an out-of-plane bending vibration of the **11** type (Figure 2.15) is negative in the difference IR-LD spectrum because the corresponding resulting transition moment is perpendicularly dispose to the plane of the benzene ring and to main symmetry axis.

The interpretation and assignment of the IR bands belonging to vibrations with mutually perpendicular resulting transition moments are especially successful for analysis using the RDP.

A comparison analysis could be performed, using 4-cyano phenyl benzoates with R = -OC₄H₉, R = -OC₆H₁₃, R = -OC₅H₁₁, and R = -OC₇H₁₅, with the chemical diagram depicted in Figure 2.17 as model compounds. These are good examples of systems with two benzene systems, which represent a more difficult case, than the liquid crystal NP-1083.

The IR spectra of these liquid crystals are characterized with series of IR bands of the substituted benzene rings, as well as the C≡N and C=O functions (Figure 2.18).

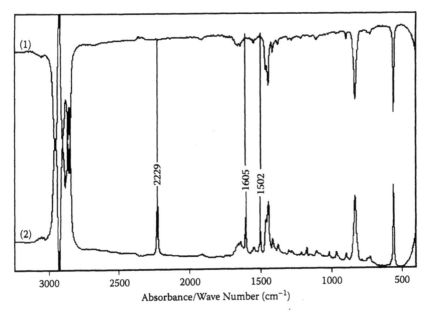

**FIGURE 2.16** (1) Reduced IR-LD of NLC NP-1083 after the elimination of the band at 1605 cm⁻¹ and (2) the corresponding nonpolarized IR spectrum.

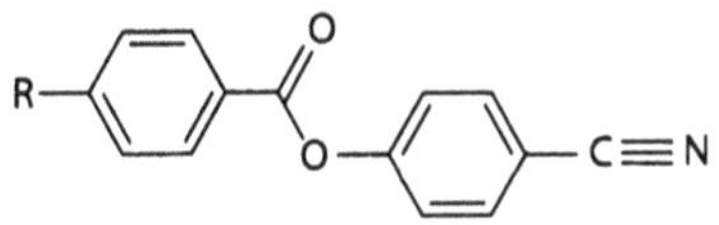

**FIGURE 2.17** Chemical diagram of the 4-cyano phenyl benzoates with R = -OC$_4$H$_9$, R = -OC$_6$H$_{13}$, R = -OC$_5$H$_{11}$, and R = -OC$_7$H$_{15}$.

For all of the liquid crystals depicted in Figure 2.17 an excellent degree of macroorientation in the solid state is obtained, which can be seen in the parallel and perpendicular IR-LD spectra shown in Figure 2.19.

The difference IR-LD spectra (Figure 2.20) show a positive IR band for the $\nu_{CN}$ stretching vibration, coinciding with the main symmetry axis of the molecules (Figure 2.17) as well as with the main crystallographic axes (Figure 2.21). This is a typical phenomenon for compounds of this type [128,129].

If the reducing difference procedure is applied on the out-of-plane IR bands (Figure 2.22), a total disappearance of the subcomponent absorption maxima at 878 cm⁻¹ and 838 cm⁻¹ of B$_1$ symmetry class (Figure 2.15) is observed. According to the Bethe theory, subsequently described as Bethe splitting (BS), the IR bands are split in the solid-state into submaxima, whose number is equal to the number of the molecules in the unit cell [128]. In the crystal of the 4′-cyanophenyl-4-n-pentylbenzoate [128], the molecule is flat and the two molecules in the unit cell are mutually disposed in a coplanar manner, thus leading to a colinear orientation of the

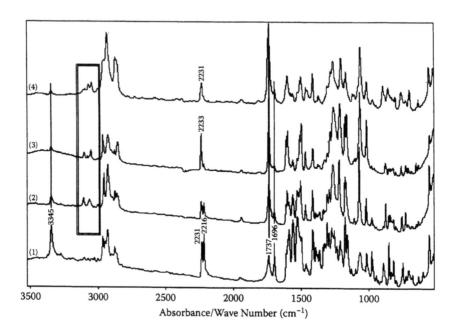

**FIGURE 2.18** IR spectra of the 4-cyano phenyl benzoates with (1) R = -OC₄H₉, (2) R = -OC₆H₁₃, (3) R = -OC₅H₁₁, and (4) R = -OC₇H₁₅.

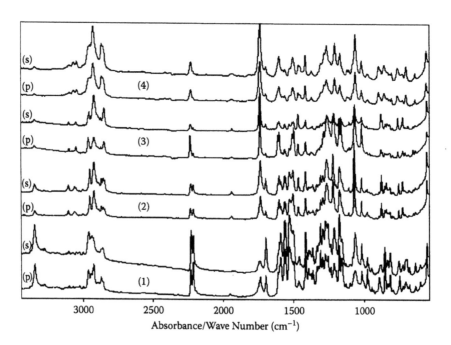

**FIGURE 2.19** Parallel (*p*) and perpendicular (*s*) IR-LD spectra of the 4-cyano phenyl benzoates with (1) R = -OC₄H₉, (2) R = -OC₆H₁₃, (3) R = -OC₅H₁₁, and (4) R = -OC₇H₁₅.

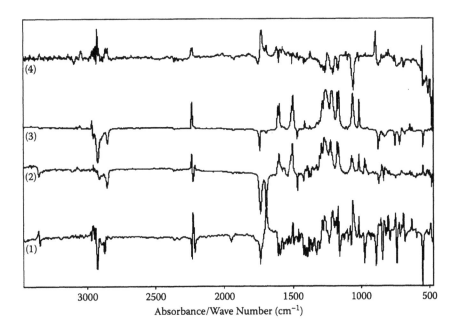

**FIGURE 2.20** Differential IR-LD spectra of the 4-cyano phenyl benzoates with (1) R = -OC$_4$H$_9$, (2) R = -OC$_6$H$_{13}$, (3) R = -OC$_5$H$_{11}$, and (4) R = -OC$_7$H$_{15}$.

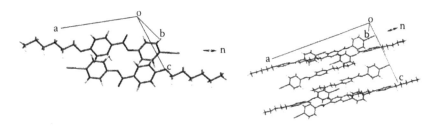

**FIGURE 2.21** Unit cells of the 4′-cyanophenyl-4-n-pentylbenzoate, 4′-cyanophenyl-4-n-heptylbenzoate [128,129], and orientation director **n**, coinciding with the main **c** axis.

transition moments of the B$_1$ symmetry class (Figure 2.23). In the reduced IR-LD spectra for these systems the subcomponents are eliminated at an equal dichroic ratio (Figure 2.22) [128].

In contrast the elimination of the $v_{C=O}$ stretching vibration of the reduced IR-LD spectrum of 4′-cyanophenyl-4-n-heptylbenzoate leads to elimination of the IR bands of out-of-plane mode at 888 cm$^{-1}$ (Figure 2.24). This result indicates a nonplanarity of the molecule, which is confirmed by the crystallographic data. The unit cell consists of Z = 8 (Figure 2.21) and pairs of nonequivalent molecules are observed, where both the benzene planes are inclined at a mutual angle of 57.7(5)° (Figure 2.25) [129]. The resulting transition moments of the IR bands of B$_1$ and $v_{C=O}$ in the framework of the unit cell result in the disappearance of the corresponding bands at an equal dichroic

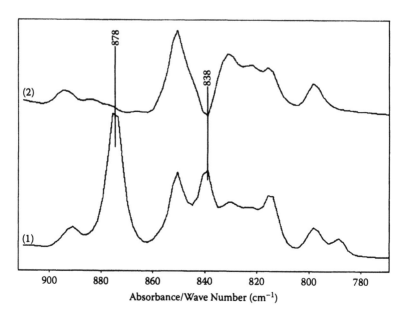

**FIGURE 2.22** (1) Nonpolarized IR and (2) reduced IR-LD spectra of 4′-cyanophenyl-4-n-pentylbenzoate after the elimination of the band at 878 cm⁻¹.

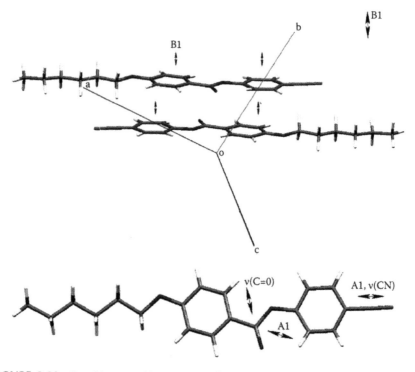

**FIGURE 2.23** Resulting transition moment direction of the B1 modes in the crystal of 4′-cyanophenyl-4-n-pentylbenzoate; visualized transition moments in the isolated molecule.

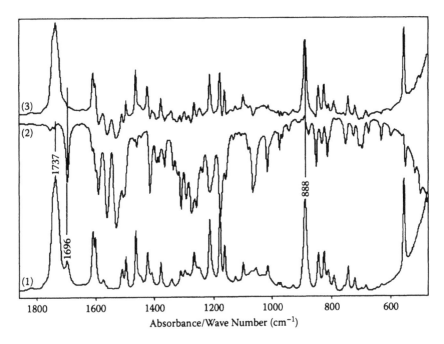

**FIGURE 2.24** (1) Nonpolarized IR and reduced IR-LD spectra of 4'-cyanophenyl-4-n-heptylbenzoate after the elimination of the bands at (2) 1737 cm$^{-1}$ and (3) 1696 cm$^{-1}$, respectively.

**FIGURE 2.25** Planes of the benzene rings in the molecule of 4'-cyanophenyl-4-n-heptylbenzoate. (Lokanath, N.K., D. Revannasiddaiah, M.A. Sridhar, J.S. Prasad, and D.K. Gowda, 1997, *Z. Kristallogr,* 212:385–339. With permission.)

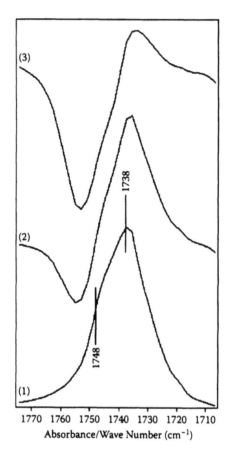

**FIGURE 2.26** (1) Nonpolarized IR and reduced IR-LD spectra of 4′-cyanophenyl-4-n-heptylbenzoate after the elimination of the bands at (2) 1748 cm$^{-1}$ and (3) 1738 cm$^{-1}$, respectively.

ratio. The elimination of the band at 1696 cm$^{-1}$ leads to disappearance of a band at 848 cm$^{-1}$, thus experimentally proving the character of the first IR maximum as an overtone.

The presence of more than one molecule in the unit cell oriented in a different way leads to splitting of the IR band according the Bethe theory and the submaxima possess different polarization. In these cases the elimination is obtained to an "inflex point" as shown in Figure 2.26, where the consequent elimination of the submaxima at 1738 cm$^{-1}$ in 4′-cyanophenyl-4-n-heptylbenzoate is obtained.

# 3 Structural Elucidation of Inorganic Compounds and Glasses

The application of the CS-NLC method for structural and local structural elucidation has been demonstrated on inorganic compounds and glasses. Crystalline oxides and borates $\alpha$-$TeO_2$, $V_2O_5$, $MoO_3.H_2O$ and its high temperature form, $\beta$-$BaB_2O_4$, $SrB_4O_7$, $H_3BO_3$, and $B_2O_3$ have been described. These systems have been the object of intensive investigations for the last 50 years, due to their variety of applications in technology, telecommunications, as NLO materials, and for fundamental knowledge. Our questions were: What can we add to the known vibrational assignments of these systems? Can we apply our orientation method for structural and local structural elucidation in these types of compounds? What are the advantages and what are the limits [130]? It must be stressed that these high-temperature melting compounds cannot be fully analyzed by the orientation method as colloid suspensions on its own. The application of the single-crystal X-ray diffraction method for inorganic compounds is possible and in these cases a correlation between the structure infrared (IR) spectra interpretation is obtained by comparing the available crystallographic data of the structure and the linear-dichroic infrared IR-LD spectra. However, for glasses, X-ray diffraction is impossible to apply and the CS-NLC method also appears to be unique for local structural determination and spectroscopic properties.

## 3.1 INORGANIC COMPOUNDS

The crystalline model systems $\alpha$-$TeO_2$, $V_2O_5$, $MoO_3.H_2O$, the high-temperature form of $MoO_3$, $\beta$-$BaB_2O_4$, $SrB_4O_7$, $H_3BO_3$, $B_2O_3$ were studied with the goal of assigning the characteristic IR bands to corresponding normal modes of the typical local structures in tellurite and borate glasses. In addition, the comparison with crystallographic data for these systems demonstrate the applicability of the CS-NLC method for structural elucidation of inorganic compounds. As model systems of glasses, we have been studying four $TeO_2$ and borate glasses as well as two binary glasses ($TeO_2$-$V_2O_5$ and $TeO_2$-$MoO_3$). The interest has focused on the glasses and crystalline inorganic chemistry (such as $\beta$-$BaB_2O_4$, $SrB_4O_7$, $MoO_3$, etc.).

As a first step we were surprised that the studied crystalline compounds were characterized by a significant degree of macroorientation of their crystalline samples (Figure 3.1), enabling an adequate interpretation of the polarized IR-LD spectra. In the $\alpha$-$TeO_2$ structure, $TeO_4$ units constitute a three-dimensional (3D) network [131] and from an IR spectroscopic point of view, $TeO_4$ groups in $\alpha$-$TeO_2$ could

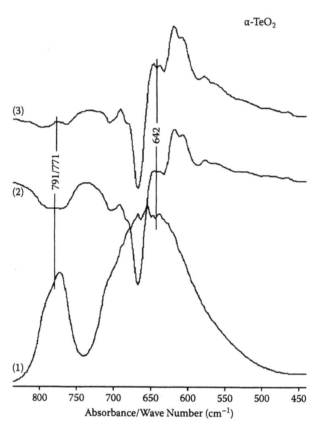

**FIGURE 3.1** (1) Nonpolarized IR, (2) difference, and (3) reduced IR-LD spectra α-TeO$_2$ after elimination of the band at 791/771 cm$^{-1}$.

be described as displaying C$_{2v}$ point symmetry and four vibrational bands were characteristic: $\nu_1$ (A$_1$, $\nu^s_{TeO2eq}$), $\nu_8$ (B$_1$, $\nu^{as}_{TeO2eq}$), $\nu_6$ (B$_2$, $\nu^{as}_{TeO2ax}$), and $\nu_2$ (A$_1$, $\nu^s_{TeO2ax}$), which were observed at 791/771 cm$^{-1}$, 707 cm$^{-1}$, 678 cm$^{-1}$, and 642 cm$^{-1}$, respectively (Figure 3.1). According to this description, the elimination of the $\nu_1$ band will lead to the disappearance of the $\nu_6$ band because they are from the same class of symmetry (Figure 3.1) and accordingly, the disappearance of the 791/771 cm$^{-1}$ and 642 cm$^{-1}$ bands is obtained at an equal dichroic ratio. These data correlated well with those reported for tellurite glasses and crystalline TeO$_2$ phases [132–134].

For pure V$_2$O$_5$ (Figure 3.2), with a distance between individual layers in the structure of 4.023 Å, an increase in the symmetry of the polyhedra is involved and in the five coordinated VO$_5$ units, an isolated V=O bond is characteristic with an intensive IR band of about 1020 cm$^{-1}$ (stretching $\nu_{V=O}$ mode) [135–140]. In the difference IR-LD spectrum of V$_2$O$_5$, multiple bands are detected, which could be explained with a deviation from the perfect VO$_5$ structure. The geometrical parameters of two neighboring VO$_5$ polyhedra are distorted [135]. The $\nu_{V=O}$ maximum is negatively oriented (Figure 3.2), on assuming an orientation of the layers along the

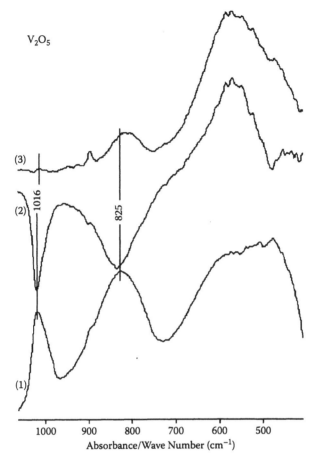

**FIGURE 3.2** (1) Nonpolarized IR, (2) difference, and (3) reduced IR-LD spectra of $V_2O_5$ after elimination of the band at 1016 cm$^{-1}$.

b axis. The elimination of the maximum at 1016 cm$^{-1}$ leads to a strong reduction of the band at 825 cm$^{-1}$; the former band belongs to the symmetric stretching V-O-V mode. The band at 870 cm$^{-1}$ is affected at a significant level and could be assigned to the asymmetric stretching mode of V-O-V.

In $\alpha$-MoO$_3$.H$_2$O, the Mo=O stretch mode lies within the 966 to 998 cm$^{-1}$ range [141,142]. A strong absorption doublet peak at 988/987 cm$^{-1}$ (Figure 3.3) in $\alpha$-MoO$_3$. H$_2$O confirms the presence of terminal double bonds, which is a basic characteristic of a layered structure. In the case of asymmetric Mo-O bonds, two resonance absorption peaks correspond to asymmetric Mo-O bonds (863 cm$^{-1}$ and 815 cm$^{-1}$). The observation of a single absorption band at 786 cm$^{-1}$ implies that Mo-O bond lengths on both sides of the O atom in the Mo-O-Mo bond are symmetric. The bending mode of the Mo-O-Mo vibration entity is between 664 cm$^{-1}$ and 468 cm$^{-1}$ (in our case at 554 cm$^{-1}$). The bands at 894 cm$^{-1}$, 725 cm$^{-1}$, 624 cm$^{-1}$, and 480 cm$^{-1}$ can be assigned to bending vibrations of the H$_2$O molecule incorporated in a different

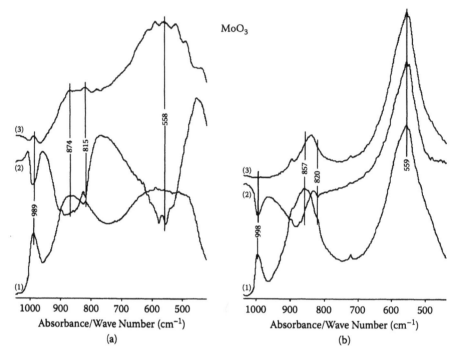

**FIGURE 3.3** (a): (1) Nonpolarized IR, (2) difference, and (3) reduced IR-LD spectra of α-MoO₃.H₂O after elimination of the band at 989 cm⁻¹. (b): (1) Nonpolarized IR, (2) difference, and (3) reduced IR-LD spectra of the high temperature form of MoO₃ after elimination of the band at 998 cm⁻¹.

manner between the layers of $MoO_3$ [143]. The IR spectrum of the high temperature form of $MoO_3$ is characterized with the absorption maxima at 998 cm⁻¹, 896 cm⁻¹, 860 cm⁻¹, 821 cm⁻¹, and 563 cm⁻¹ (Figure 3.4). The IR characteristic bands confirm the layer structure of the high temperature form due to the typical $v_{Mo=O}$ band [141,142] at 998 cm⁻¹.

Similar to α-$MoO_3$.H₂O, a significant degree of macroorientation [33–35] of suspended particles is obtained (Figure 3.3a,b). The existence of a layered structure in both $MoO_3$ forms is confirmed from the reducing IR-LD spectra, where the elimination of the bands at 989 cm⁻¹ or 998 cm⁻¹ leads to a disappearance of a maximum at about 817 cm⁻¹. These results experimentally clarify the character of the 817 cm⁻¹ band, which is commonly an asymmetric stretching mode. However, its reduction at an equal dichroic ratio with the $v_{Mo=O}$ peak indicates a symmetric stretch origin, due to the colinear transition moments in the framework of the $MoO_3$ layered structure.

Boroxol rings BO₃ and BO₄ are proposed [144–148] as characteristic local fragments (Figure 3.4) in borate glasses. The model systems of crystalline borate compounds with known crystallographic structures and the available structural information give the possibility for adequate assignments of IR peaks to corresponding normal modes, characteristic of given local structural units in the glasses.

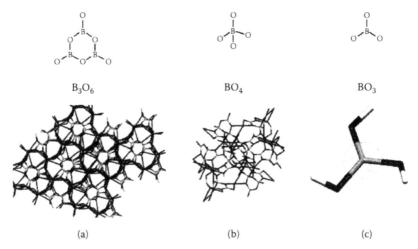

$B_3O_6$         $BO_4$         $BO_3$

(a)         (b)         (c)

**FIGURE 3.4**  Possible local structures in borate glasses [130] as well as the crystal structures of (a) $\beta$-BaB$_2$O$_4$, (b) SrB$_4$O$_7$, and (c) H$_3$BO$_3$.

The crystal structure of $\beta$-BaB$_2$O$_4$, which has been intensively studied due to its NLO application, consists predominantly of boroxol units (Figure 3.4) [149,150]. The obtained geometry parameters show that the interactions in the solid state affect the B$_3$O$_6$ structural units leading to their distortion from planarity. The observed bond lengths and angles of 1.411 and 1.403 Å and 123.8(8), 121.2(3) and 94.1(0)° suggest that multiple IR absorption bands will be characteristic for these moieties.

The positives in the difference IR-LD spectrum bands of $\beta$-BaB$_2$O$_4$ at 1492 cm$^{-1}$, 770 cm$^{-1}$, and 635 cm$^{-1}$ are eliminated at equal dichroic ratios, confirming the A$_1'$ symmetry class [149]. The negative bands at 1302 cm$^{-1}$ and 480 cm$^{-1}$ are also eliminated simultaneously, confirming their assignment to the E$'$ symmetry class [58]. The selected sets of bands and reduction results can be used as characteristics for boroxol rings in the IR spectra of borate glasses or for other inorganic derivatives. Typically, for the boron oxide structure the Raman active band at 808 cm$^{-1}$ has a very low intensity in IR spectra [147], and for this reason is not characteristic and cannot be employed as a characteristic for the presence or absence of boroxol units.

As a model system of borate compound with BO$_3$ and BO$_4$ structural units, we studied SrB$_4$O$_7$ (Figure 3.4) [151]. Similar to $\beta$-BaB$_2$O$_4$, the interactions in the condensed phase result from a distortion of BO$_3$ and BO$_4$, reflecting the observed multiple characters of the IR bands. Independently, the application of the reducing difference procedure for polarized IR-LD spectra show that: (a) the elimination of the bands at 642 cm$^{-1}$, 628 cm$^{-1}$, 811 cm$^{-1}$, and 779 cm$^{-1}$ at equal dichroic ratios confirms that they are in the F$_2$ symmetry class in the BO$_4$ unit; and (b) the reduction of both broad bands at 1249 cm$^{-1}$ and 1159 cm$^{-1}$ at equal dichroic ratios confirms the $\nu_{B-O}$ stretches of BO$_3$, where the obtained values are similar to those reported by Stark [48]. The IR band for $\nu_4$ (E$_1$), observed at 555 cm$^{-1}$, is typical for BO$_3$ [148,149]. However, in the glasses it is of low intensity and difficult to assign. The discussed selected set of bands for BO$_3$ fragments is confirmed on studying crystalline H$_3$BO$_3$ [152]. In the framework of the unit cell the $B(OH)_3$ moieties form layered two-dimensional (2D) structures, which are practically mutually parallel. Similar to the IR-LD spectrum of $SrB_4O_7$, the elimination of the bands at 821 $cm^{-1}$ and 649 $cm^{-1}$ at equal dichroic ratios (Figure 3.5) confirms the previously established assignment. The $v_4$ ($E_1$) band at 547 $cm^{-1}$ is practically unaffected.

Using these sets of IR bands, the interpretation of the IR-LD spectrum of pure $B_2O_3$ was performed. The IR spectroscopic bands are well defined starting from the 1800 $cm^{-1}$ region, which has been previously reported [97]. According to the crystallographic data of crystalline $B_2O_3$, the main structural unit in this case is $BO_3$, however, the IR spectroscopic patterns within the whole 1800 to 400 $cm^{-1}$ IR region are complicated at a significant level and are different from the simpler IR spectra of inorganic compounds with $XY_3$ units [58]. Their multiple characters could be explained with distorted polyhedra in the unit cell, but the observation of the maxima higher than 1600 $cm^{-1}$ are typical for ring systems. On the other hand, a comparison between the IR spectra of crystalline $B_2O_3$ and $\beta$-$BaB_2O_4$ shows differences, thus speaking against boroxol rings as a structural unit. The data suggest the formation of supramolecular rings, which is confirmed by the elimination of the bands at 1515 $cm^{-1}$, 790 $cm^{-1}$, and 640 $cm^{-1}$ at the same dichroic ratios. Additional

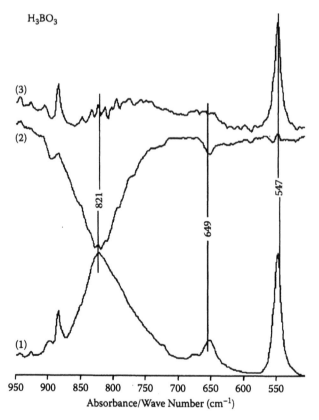

**FIGURE 3.5** (1) Nonpolarized IR, (2) difference, and (3) reduced IR-LD spectra of $H_3BO_3$ after elimination of the band at 821 $cm^{-1}$.

confirmation of the assumptions presented earlier is provided by the simultaneous disappearance of the 1800 to 1600 cm⁻¹ bands indicative of combination modes or overtones. These data correlated well with the crystallographic data for $B_2O_3$ [153,154] illustrating the applicability of the CS-NLC method for an adequate analysis of the correlation between the structure and the IR spectroscopic properties.

## 3.2  IR SPECTROSCOPIC ELUCIDATION OF GLASSES

In the difference IR-LD spectrum of $TeO_2$ *glasses G1* and *G2* (Figure 3.6) a well-defined degree of macroorientation of the suspended particles is observed. This result follows the important conclusion that in the glasses a long-range order partially exists. Changes of the intensity bands, can be explained by the distortion of the $TeO_4$ polyhedra in the *glass G1* [137,149]. They are predominantly of the $v^s_{TeO2eq}$ type in the IR spectrum of the glass in contrast to the analogous one of $\alpha$-$TeO_2$. The elimination of the bands at 773 cm⁻¹ and 642 cm⁻¹ at equal dichroic ratios confirms their assignment to the $A_1$ symmetry class (Figure 3.6). The IR-LD spectroscopic study of $TeO_2$ *glass G2* shows that the bands at 773 cm⁻¹ and 660 cm⁻¹ are eliminated at the same dichroic ratio, suggesting a predominant presence of $TeO_4$ local structural units. However, the obtained reduced IR-LD spectrum is characterized by bands at 615 cm⁻¹ and 705 cm⁻¹, which are typical for distorted $TeO_3$ trigonal pyramids moieties as well.

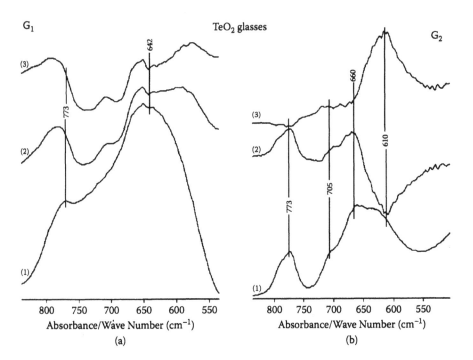

**FIGURE 3.6**   (a): (1) Nonpolarized IR, (2) difference, and (3) reduced IR-LD spectra of $TeO_2$ *G1*, after elimination of the band at 773 cm⁻¹. (b): (1) Nonpolarized IR, (2) difference, and (3) reduced IR-LD spectra of $TeO_2$ *G2*, after elimination of the band at 773 cm⁻¹.

The nonpolarized and difference IR-LD spectra of $TeO_2$-$V_2O_5$ indicate a relatively high degree of macroorientation of suspended particles, which suggests a long-range order in the system. The obtained low-frequency shifting of $v_{V=O}$ band in the glasses, compared with crystalline $V_2O_5$ (45 cm$^{-1}$ in our case) is typical for $V_2O_5$ glasses and has been systematically studied [137]. Similar to pure $V_2O_5$ the elimination of the last band and the disappearance of the maximum observed at 825 cm$^{-1}$ indicate the presence of $VO_5$ units in the studied system. On the other hand, the reduction of the bands at 797 cm$^{-1}$ and 668 cm$^{-1}$ at the same dichroic ratio indicates the presence of $TeO_4$ units. The remaining bands at 602 cm$^{-1}$ and 718 cm$^{-1}$, like in the case of the $TeO_2$ glass *G2*, could be attributed to $TeO_3$ units.

The polarized IR-LD spectroscopic study of $TeO_2$-$MoO_3$ glass is also in accordance with the assumption of the presence of a long-range order in the studied glasses. The degree of macroorientation of the solid sample is reasonable. The difference spectrum shows bands at 991 cm$^{-1}$ and 791 cm$^{-1}$, indicating that deformed $[MoO_6]$ octahedral units are present in the glass. The bands at 723 cm$^{-1}$ and 585 cm$^{-1}$ could be explained by the presence of $TeO_3$ units. This data correlates with that of previously published investigations [153–155].

The IR spectra of borate *glasses G3* and *G4* are illustrated in Figure 3.7. The elimination of the intensive bands at 1240 cm$^{-1}$, 1155 cm$^{-1}$, and 555 cm$^{-1}$ in the difference IR-LD spectrum (Figure 3.7) is explained by the presence of $BO_3$ local

**Borate Glasses**

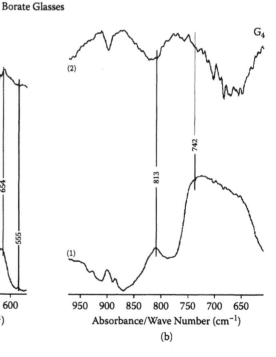

FIGURE 3.7   (a): (1) Nonpolarized IR and (2) difference IR-LD spectrum of borate *glass G3*. (b): (1) Nonpolarized IR and (b): reduced IR-LD spectrum of borate *glass G4* after elimination of the band at 813 cm$^{-1}$.

structures in *G3*. However, the elimination of the 1500 cm$^{-1}$, 760 cm$^{-1}$, and 653 cm$^{-1}$ peaks at equal dichroic ratios suggests the presence of boroxol ring structures as well. In contrast, the IR-LD spectra of *glass G4* shows the simultaneous disappearance of the selected set of bands at 813 cm$^{-1}$ and 742 cm$^{-1}$ indicating the presence of $BO_3$ fragments in *G4*. However, the presence of the bands at 3400 cm$^{-1}$ and 1630 cm$^{-1}$ is associated with the absorption of the water molecule and the "parasite" presence of $H_3BO_3$. Moreover, some of the characteristic bands for the latter compound ($H_3BO_3$) are observed as well.

# 4 Structural Elucidation of Organic Compounds

More than 350 organic compounds in different classes have been investigated by means of the CS-NLC method. It has been applied on heterocyclics, including substituted aminopyridines, quinolines, azaindoles, their acetylated derivatives [156–164]; benzimidazoles, benzotriazoles, their salts, and some metal complexes [165–167]; salts and ester amides of all essential amino acids; small aliphatic and aromatic peptides, their protonated forms, and metal complexes [168–203]; polymorph modifications of pharmaceutical products such as paracetamol and aspirin, some analgetics, cephalosporin, and penicillin antibiotics [204–207]; morphine alkaloids [208–211]; organic dyes [212–218]; and other classes of organic compounds such as lactams, benzamidines, squaric, croconic, and rhodizonic acids [219–223]. The data for substituted amino acids and small peptides have been summarized in book chapters and a review article [224–226]. In this book we present only a small part of these investigations with the goal to illustrate the scopes and limitations of the linear-dichroic infrared (IR-LD) method for these systems. The effects in the IR spectra of crystals are elucidated using series of derivatives of squaric and violuric acids [227,228].

## 4.1 ANALYSIS OF HETEROCYCLIC COMPOUNDS

In this section we will demonstrate the applicability of the CS-NLC method for investigation of the effect of protonation on the symmetry of the molecules and the IR spectroscopic properties. In this respect, the CS-NLC method has also been successfully applied for the elucidation of organic dyes, in particular different dicyanoisophorones, anyles, merocyanine dyes, and corresponding stilbazolium salts.

The distortion of aromatic character as a result of protonation in different substituted pyridines (i.e., 4-aminopyridine, 4-dimethylaminopyridine, 3,4-diaminopyridine, and 4-cyanopyridine) has been elucidated systematically [157,160,161,163,164]. The IR spectroscopic data have been compared, supported, or confirmed by UV- and [1]H-NMR-spectral analysis, hybrid methods HPLC/ESI MS-MS, and single crystal X-ray diffraction (XRD). The model systems 4-aminopyridinium palladate(II), 3,4-diaminopyridinium hydrogentartarate, 3,4-diaminopyridinium bis(perchlorate), dimethylaminopyridinium hydrogensquare, and 4-cyanopyridinium hydrogensquarate monohydrate will be described here. During these investigations a new phenomenon has been identified, which is also observed in the violurate salts of heterocyclic compounds [228]. The red color of 4-cyanopyridinium hydrogensquarate monohydrate in the solid state is a result of the strong bathochromic shifting of the CT band within the framework of the stable layers formed by $(2HSq^-.2H_2O)_n$ fragments. These represent a new structural motif for squaric acid anions. Quantum chemical calculations on the electronic structure and vibrational analysis at the Moeller-Plesset (MP2), restricted Hartree–Fock (RHF), or density functional theory (DFT) level of theory and high basis sets of 6-31++G** or 6-311++G** types have been presented [164].

Commonly, the interest in different substituted aminopyridines is based on the biological activity of the neutral compounds and more likely of their protonated forms as $K^+$ and $Ca^{2+}$ channel inhibitors [229]. In the latter forms the calcium influx is enhanced, leading to an increase in the release of neurotransmitter [230]. Due to their ability to facilitate nerve transmission, substituted aminopyridines have been applied to reverse anaesthesia and muscle relaxation. They have been proposed as drugs for the treatment of multiple sclerosis, myasthenia gravis, spinal cord injuries, botulism, and Alzheimer's disease. Aminopyridines have a p$Ka$ ~ 9.0, and thus they can exist in both the neutral or cationic protonated form at physiological pH. This characteristic complicates the elucidation of the mechanism and site of action [229–232]. Theoretical studies of the molecular determinants responsible for the biological activity of aminopyridines have been presented [233] at the B3LYP/cc-pVDZ level of theory. The results have confirmed pyramidal and planar structures for the neutral form and cations, respectively. Assuming a biological effect of the 4-aminopyridine complexes, their structures and more commonly the 4-aminopyridine coordination capability have also been studied [234]. There is sound evidence that this compound acts monodentately as a result of a considerable strong nucleophilic ability of the pyridine (endocyclic) nitrogen. Such coordination should not change the aromatic structure of the ligand and this is proven by comparison of the pyridine skeleton characteristic bands in the IR spectra of 4-aminopyridine and its complexes with transition metals. However, a different situation with respect to the 4-aminopyridine salts has been suggested. The theoretical results including STO-3G* *ab initio* calculations also predict the endocyclic nitrogen as a favorable position of protonation, but a final stabilization of the "imino" form has been established using an IR spectral study of 4-aminopyridine hydrochloride. Similar conclusions have been obtained by means of a comparative nuclear magnetic resonance (NMR) spectral analysis of 4-aminopyridine and its salts. Interestingly, this phenomenon has not been discussed in articles investigating the crystal structures of 4-aminopyridine salts. This is probably because X-ray analysis can determine the proton location but it does not present reliable information about the electronic distribution. Despite the fact that protonation processes are of key importance, crystallographic data especially for mono- and diprotonated 3,4-diaminopyridine is limited. Crystal structures of tetrakis(2-Aminopyridinium) 2-ammoniopyridinium bis(μ-5-phosphato)-pentakis(μ-2-oxo)-decaoxo-penta-molybdenum [235], the tetrachloro- and tetrabromocuprate(II) salts of 3-aminopyridine [236], and 4-ammoniopyridinium tris(μ-2-chloro)-hexachloro-dimolybdenum monoethanolamine are known. The diprotonated form of 3,4-diaminopyridine as a bis(perchlorate) was reported for the first time by Koleva et al. [161].

A main point of emphasis of these investigations is the possibility of a comparison between the IR-LD spectroscopic data and those obtained by single-crystal XRD.

The salt bis(4-aminopyridinium) tetrachloropalladate(II) is formed by a discrete $PdCl_4^{2-}$ anion and two $N_{py}$ protonated $C_5H_7N_2^+$ cations without appreciable Pd···N interactions (Figure 4.1). The geometry of Pd(II) ion is square planar (Pd-Cl(1),

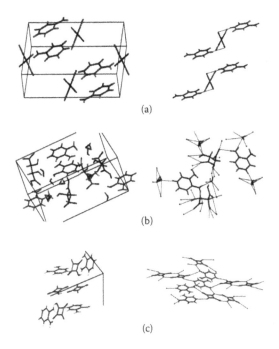

(a)

(b)

(c)

**FIGURE 4.1** Unit cells and hydrogen bonding of (a) bis(4-aminopyridinium) tetrachloropalladate(II), (b) 3,4-diaminopyridinium hydrodentartarate dihydrate, and (c) 3,4-diaminopyridinium hydrogensquarate.

Pd-Cl(2) bond lengths are 2.2940(11)Å and 2.3073(14)Å, respectively, and Cl-Pd-Cl angles are 180.00°, 89.99(5)°, and 90.01(5)°).

3,4-Diaminopyridinium hydrogentartarate dihydrate crystallizes in the noncentrosymmetric space group $P2_12_12_1$. The molecules form a 3D network (Figure 4.1) by intermolecular NH···OH$_2$ (2.807 Å), HNH···OH$_2$ (3.037 Å), HNH···OCO (3.007 Å), HOH···OCO (2.680 Å), OH···OH$_2$ (2.789 Å), and HOH···OH (2.747 Å) hydrogen bonds. Hydrogentartarate anions form chains through strong intermolecular O=COH···OCO interaction (2.517 Å). It is interesting to observe that only the 4-NH$_2$ group of the cation takes part in an intermolecular interaction, which is in contrast to neutral 3,4-diaminopyridine. A similar pattern is exhibited in the corresponding hydrogensquarate salt [227], where only the NH and 4-NH$_2$ groups are joined in the intermolecular NH···O=C$_{(Sq)}$ (2.800 Å) and NH$_2$···O=C$_{(Sq)}$ (2.969 and 2.974 Å) interactions (Figure 4.1). The molecular cation is flat with a maximal deviation from total planarity less than 0.4°. The NH and NH$_2$ groups lie in the plane of the pyridine ring with a deviation of 2.02°.

3,4-diaminopyridinium bis(perchlorate) crystallizes in the noncentrosymmetric space group $Cc$. 3,4-Diaminopyridinium cations and ClO$_4^-$ anions form infinite chains by moderate intermolecular interactions of types NH$_3^+$···OClO$_3^-$ with bond lengths of 3.031, 3.024, 2.825, and 2.875 Å (Figure 4.2). The pyridine NH group participates in intermolecular contacts NH···OClO$_3^-$ with bond lengths of 3.220 and 3.172 Å, respectively. Similar to the previously studied salts [174], only the 4-substituent participates in interactions. This is in contrast to neutral 3,4-diaminopyridine [237], where both 3- and 4-$NH_2$ substituents participate in symmetric intermolecular hydrogen bonds (HNH⋯N, 2.258, 2.228 Å (4-$NH_2$) and 2.327, 2.496 Å (for 3-$NH_2$)). The geometry of the cation is effectively flat with a maximal deviation from total planarity less than 0.2°. The NH and 3-$NH_2$ groups lie in the plane of the pyridine ring with a deviation of 1.07°. These data are in accordance with the theoretically predicted geometric parameters for the dication of 3,4-diaminopyridine where differences of less than 0.031 Å and 2.6° are obtained.

The 4-dimethylaminopyridinium cation interacts through the NH⋯O bonds with the hydrogensquarate dimers (bond lengths and angles of 2.729 Å and 165.0(0)°) in the corresponding salt (Figure 4.1). The resulting subunits are π–π stacked with the neighboring ones (Figure 4.2b) at the distance of 3.406 Å. The cation is effectively flat with a deviation from planarity of less than 1.53°. The corresponding distances and angles for $[C_7H_{11}N_2]^+$ are typical for the protonated forms of 3,4-diaminopyridine reported in more than 38 crystallographically determined structures [237–252].

4-Cyanopyridinium hydrogensquarate monohydrate crystallizes in the space group P-1 (Figure 4.2). The molecules are joined to form infinite layers by intermolecular NH⋯$O_{(HSq^-)}$ (2.651 Å) and HOH⋯$O_{(HSq^-)}$ (2.792 and 2.563 Å) hydrogen bonds with participation of cations, anions, and solvent molecules. The formation of stable layered 2$HSq^-$.2$H_2$O fragments is obtained. Hydrogensquarates ($HSq^-$) form

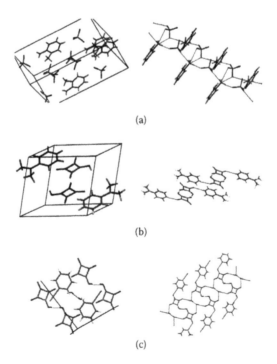

(a)

(b)

(c)

**FIGURE 4.2** Unit cell and hydrogen bonding in (a) 3,4-diaminopyridinium bis(perchlorate), (b) 4-dimethylaminopyridinium hydrogensquarate, and (c) 4-cyanopyridinium hydrogensquarate monohydrate; stable layers of repeated 2$HSq^-$.2$H_2$O fragments.

stable dimers with an equally positioned proton between the anions. The geometry of the cation is flat with a maximal deviation from total planarity less than 0.2°. The obtained bond lengths and angles are typical, on comparing the values with the data for the four known salts of 4-cyanopyridine [253–255].

The most typical IR bands of aromatic 4-aminopyridine are those of pyridine skeleton vibrations at 1602 cm$^{-1}$ and 1508 cm$^{-1}$ and the band at 824 cm$^{-1}$ of an out-of-plane (o.p.) mode, assigned according to Wilson's notation [256,257] as **8a**, **19a**, and **11**-$\gamma_{CH}$, respectively. These bands are absent in the IR spectrum of bis(4-aminopyridinium) tetrachloropalladate(II). However, three new bands at 1590 cm$^{-1}$, 1530 cm$^{-1}$, and 780 cm$^{-1}$ are observed. According to Spiner [257] and Batts and Spiner [258], the first two maxima correspond to mixed $\nu_{C=N+}$ and $\nu_{C=C}$ vibrations of the imino form. The band at 780 cm$^{-1}$, which in our case should be assigned to the $\gamma_{=CH}$ out-of-plane mode, is observed at 806 cm$^{-1}$ in 4-aminopyridine hydrochloride. The NH$_2$-stretching vibrations in the solid-state IR spectrum of 4-aminopyridine (Figure 4.3) belong to the IR bands

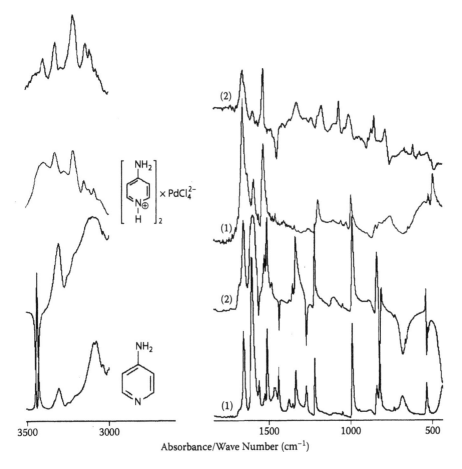

**FIGURE 4.3** (1) Nonpolarized IR and (2) difference IR-LD spectra of 4-aminopyridine and its tetrachloropalladate(II) salt.

at 3437 cm$^{-1}$ ($v^{as}_{NH2}$), 3305 cm$^{-1}$, and 3085 cm$^{-1}$ ($v^s_{NH2}$). The observation of the doublet IR band for the $v^s_{NH2}$ vibration is a result of FR, usually observed when the NH$\cdots$N$_{py}$ hydrogen bonds are formed. The corresponding IR maxima in the salt are observed at 3394 cm$^{-1}$ ($v^{as}_{NH2}$), 3325 cm$^{-1}$, and 3120 cm$^{-1}$ ($v^s_{NH2}$), respectively. In this case, the FR splitting is probably caused by (H)N$^+$-H$\cdots$Cl hydrogen bonding (H$\cdots$Cl distance of 2.58Å). The new IR band at 3215 cm$^{-1}$ is assigned to $v_{NH}$ vibration created by the protonation of the new N$_{py}$-H bond. The band of the scissoring ($\delta_{NH2}$) vibration is observed at 1650 cm$^{-1}$ in the 4-aminopyridine IR spectrum. The corresponding IR band in the tetrachloropalladate(II) salt is insignificantly high frequency shifted to 1656 cm$^{-1}$.

The aforementioned characteristics of the IR band assignment of protonated 4-aminopyridine are confirmed by the solid-state IR-LD spectral analysis. The elimination of the intensive band ($\delta_{N+H2}$) at 1645 cm$^{-1}$ provokes the disappearance of all bands belonging to molecular motions with resulting transition moments along the same direction. These are the doublet IR band of $v^s_{NH2}$ at 3323 cm$^{-1}$ and 3143 cm$^{-1}$, the maximum of the $v_{NH}$ vibration at 3222 cm$^{-1}$ as well as the IR-bands with mixed characters of $v_{C=N+}$ and $v_{C=C}^-$ vibrations at 1632 cm$^{-1}$, 1588 cm$^{-1}$, and 1532 cm$^{-1}$ (Figure 4.3).

The solid-state IR spectrum of neutral 3,4-diaminopyridine (3,4-DAP) within the 3500 to 3000 cm$^{-1}$ IR region is characterized by four bands at 3390 cm$^{-1}$ ($v^{as}_{NH2}$ of 4-NH$_2$), 3351 cm$^{-1}$ ($v^{as}_{NH2}$ of 3-NH$_2$), 3281 cm$^{-1}$ ($v^s_{NH2}$ of 4-NH$_2$), and 3193 cm$^{-1}$ ($v^s_{NH2}$ of 3-NH$_2$). FR splitting of the IR band of $v^s_{NH2}$ is absent as a result of the symmetric hydrogen bond interaction of both NH$_2$ groups [259]. According to the crystallographic data both NH$_2$ groups take part in the NH$_2\cdots$N contacts (3.145 to 3.184 Å). The bands at 1670 cm$^{-1}$ and 1622 cm$^{-1}$ belong to $\delta_{NH2}$ of the 4- and 3-substituents. The consequent elimination of the last maxima leads to the disappearance of 3281 cm$^{-1}$ and 3193 cm$^{-1}$ at equal dichroic ratios (Figure 4.4), which is a result of the colinear orientation of corresponding transition moments in the plane of each of the NH$_2$ groups. The elimination of the band at 902 cm$^{-1}$ leads to the disappearance of those at 833 cm$^{-1}$ and 821 cm$^{-1}$ due to their colinear orientation in each 3,4-diaminopyridine (Figure 4.5). However, the last procedure increases the second pairs of maxima, which is attributed to the same vibrations of the second, differently oriented molecules of 3,4-diaminopyridine in the unit cell. According to crystallographic data, the molecules of the neighboring chains are disposed at an angle of 60.38° (Figure 4.6).

The IR spectrum of the 3,4-diaminopyridinium salt is more complicated than that of the neutral molecule as a result of self-absorption IR bands of solvent molecules and hydrogentartarate ions. The 3500 to 300 cm$^{-1}$ region is characterized by an overlapped broad band between 3500 and 3450 cm$^{-1}$ corresponding to the $v_{OH}$ stretch of hydrogen-bonded OH groups of solvent molecules and hydrogentartarate ions. The characteristic bands at 3490 cm$^{-1}$ ($v^{as}_{NH2}$ of 3-NH$_2$) and 3320 cm$^{-1}$ ($v^s_{NH2}$ of 3-NH$_2$) are shifted to high frequency. According to the crystallographic data, the 3-NH$_2$ group does not participate in intermolecular interactions and for this reason it is characterized by relatively high frequency shifts of its maxima. The NH$_2$ group in the 4-position forms a symmetric intermolecular hydrogen bond and is characterized by the bands at 3381 cm$^{-1}$ ($v^{as}_{NH2}$) and 3225 cm$^{-1}$ ($v^s_{NH2}$). The new band at 3270 cm$^{-1}$ is attributed to stretching motions of the NH group. The IR bands of $\delta_{NH2}$ frequencies are shifted to a lower frequency (1644 cm$^{-1}$ (4-NH$_2$) and 1613 cm$^{-1}$ (3-NH$_2$)). In the salt the intermolecular interactions with participation of the 4-NH$_2$

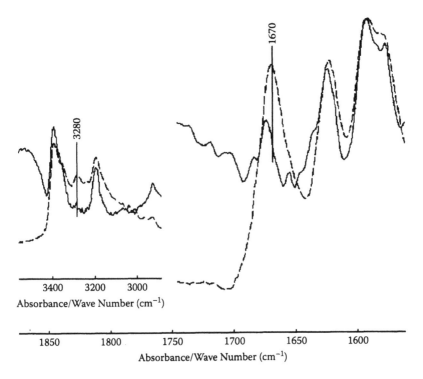

**FIGURE 4.4** Nonpolarized IR (dashed line) and reduced IR-LD (solid line) spectra of 3,4-diaminopyridine after elimination of the band at 1670 cm⁻¹.

group are weaker and its $\delta_{NH2}$ has a difference of 26 cm⁻¹ compared to the value of 3,4-diaminopyridine. The significant low-frequency shifting of the bands of in-plane (i.p.) modes by less than 40 cm⁻¹ after protonation confirms the assumption of partial distortion of the aromatic character following protonation. The intensive new band at 1511 cm⁻¹ belongs to the $\delta_{NH}$ mode. In contrast to neutral molecules, the elimination of the last IR bands at equal dichroic ratios does not lead to second pairs of peaks, due to the near coplanar disposition of all the 3,4-diaminopyridinium cations in the unit cell. They exhibit an interplanar angle of 5.66°.

Like other 3,4-diaminopyridine derivatives, the difference IR-LD spectrum of the corresponding bis(perchlorate) derivative displays a significant degree of macroorientation of suspended particles. The bands at 941 cm⁻¹, 882 cm⁻¹, and 806 cm⁻¹ are eliminated in the difference IR-LD spectrum. According to the crystallographic data, the protonated 3,4-diaminopyridine moieties are disposed in a coplanar manner in the unit cell and exhibit an interplanar angle of only 1.16°. This means that the transition moment directions of the discussed o.p. maxima are disposed perpendicularly to the pyridine ring plane, that is, along the *c* axis. The elimination of corresponding bands in the difference IR-LD spectrum (Figure 4.7) indicates that their resulting transition moments adopt an angle of 54.7° for the macroorientation of the sample in a nematic host. The molecular chains oriented along the *c* axis are disposed at a magic angle toward the liquid crystal orientation director **n**. The assignment of the bands

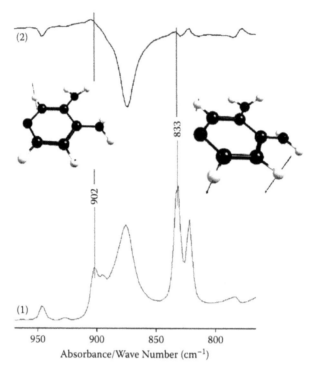

**FIGURE 4.5**  (1) Nonpolarized IR and (2) reduced IR-LD spectra of 3,4-diaminopyridine after elimination of the band at 833 cm⁻¹.

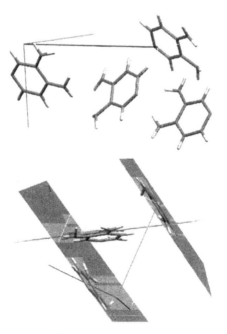

**FIGURE 4.6**  Crystal structure of 3,4-diaminopyridine.

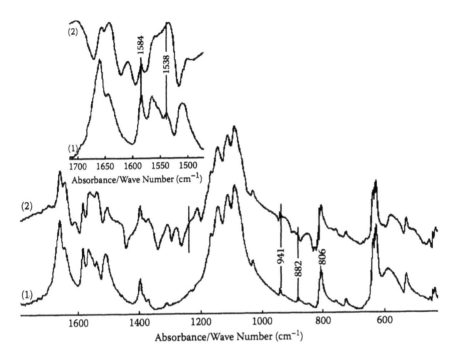

**FIGURE 4.7** Top: (1) Nonpolarized IR and (2) difference IR-LD spectra of 3,4-diaminopyridinium bis(perchlorate). Bottom: (1) Nonpolarized IR and (2) reduced IR-LD spectra after elimination of the band at 941, 882, and 806 cm⁻¹.

at 1584 cm⁻¹ and 1538 cm⁻¹ to i.p. $8_a$ and $19_a$ pyridine modes is confirmed by their simultaneous disappearance at equal dichroic ratios.

The experimental IR spectroscopic assignment of 3,4-diaminopyridine and its mono- and diprotonated forms correlated well with the corresponding theoretical vibrational analysis. The good agreement between the theoretical and experimental data is confirmed by the obtained differences of less the 7 cm⁻¹ using the unrestricted Hartree–Fock (UHF) level of theory and the 6-31++G** basis set (Figure 4.8).

The elucidation of the tautomerism in 4-dimethylaminopyridine as a result of $N_{py}$ protonation by means of polarized IR-LD spectroscopy has also been achieved. The low melting point of the neutral compound allows its investigation by means of linear-polarized IR-LD spectroscopy using the classical orientation tool as melted polycrystalline film. As can be seen, the difference IR-LD spectrum is characterized with a significant degree of orientation of the sample (Figure 4.9). The spectrum is characterized by a positive **20a** maximum at 3088 cm⁻¹ as well as maxima for **8a** and **19a** at 1627 cm⁻¹ and 1611 cm⁻¹, values typical for pyridinium derivatives. The positive sign of these bands in the difference IR-LD spectrum indicate the parallel orientation of the molecules in the crystal structure. The corresponding transition moments of the $A_1$ symmetry are oriented in a colinear fashion in the unit cell. The positive orientation of corresponding IR bands is in accordance with their orientation toward director **n** at an angle within the range of 0.0° to 54.7°. In this respect the observed negative orientation of the band assigned to $11\text{-}\gamma_{CH}$ at 812 cm⁻¹ correlates

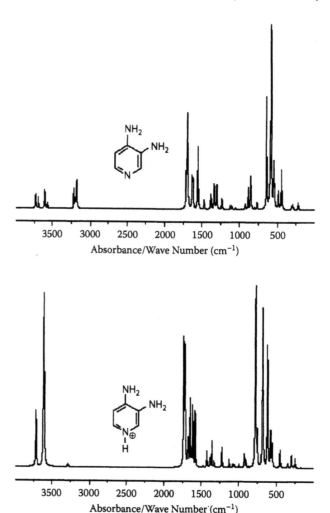

**FIGURE 4.8**   Theoretical IR spectra of non-, monoprotonated 3,4-diaminopyridine.

well with the common theory that the transition moments of $B_1$ symmetry are disposed perpendicular to the pyridine plane at an angle of 90.0°. The direct experimental confirmation of this assignment follows from the obtained reduced IR-LD
spectra, where the elimination of the discussed $A_1$ bands at equal dichroic ratios
(Figure 4.10) as well as the bands at 812 cm⁻¹ and 534 cm⁻¹ is observed. The negative
maxima within the 3000 to 2800 cm⁻¹ region as well as at 1537 cm⁻¹ (Figure 4.10)
are assigned to asymmetric and symmetric stretching ($v^{as}_{CH3}$, $v^s_{CH3}$) and bending **19b**
($B_2$ symmetry class) modes of -$CH_3$ and the pyridine ring.

The comparison of the IR spectra of the neutral 4-dimethylaminopyridine and
its monoprotonated form show low intensive peaks for the aromatic ring stretching vibrations and the observation of the series of intensive bands at 1640 cm⁻¹,
1636 cm⁻¹ and 1557 cm⁻¹, and at 873 cm⁻¹, 838 cm⁻¹, and 752 cm⁻¹ in the salt. This

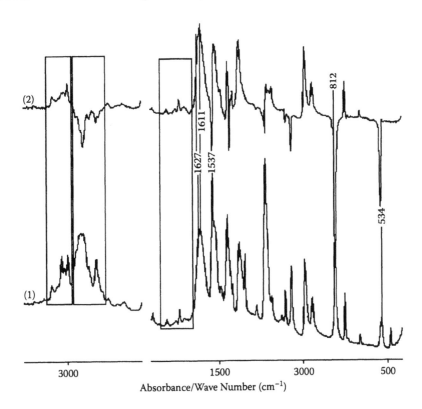

**FIGURE 4.9** (1) Nonpolarized IR and (2) difference IR-LD spectrum of 4-dimethylaminopyridine (4-DMAP).

result indicates that the $N_{py}$ protonation in solid state leads to stabilization of the imino-like form. Direct evidence follows from the reduced IR-LD spectra in an oriented colloid suspension. The consequent elimination of the bands at 1640 cm$^{-1}$ and 890 cm$^{-1}$ leads to the disappearance of corresponding maxima at 1556 cm$^{-1}$ and 730 cm$^{-1}$. Other observed IR bands at 1811 cm$^{-1}$, 1643 cm$^{-1}$, and 1565 cm$^{-1}$ are typical for $v^s_{C=O}$, $v^{as}_{C=O}$, and $v_{C=C}$ stretching vibrations of hydrogensquarate species.

Calculated IR spectra of neutral and $N_{py}$ protonated 4-cyanopyridine are shown in Figure 4.11. New IR bands of the NH$^+$-group of the pyridinium ring ($v_{NH+}$ stretching and $\delta_{NH+}$ banging modes) are observed at 3424 cm$^{-1}$ and 1592 cm$^{-1}$ of pyridinium ring and the substituent (-CN group) is only weakly affected by the $N_{py}$ protonation. The $v_{CN}$ is shifted to higher wave numbers by 4 cm$^{-1}$ after the protonation and is predicted at 2228 cm$^{-1}$. An increase in the intensity and a high-frequency shifting of the i.p. aromatic bands in the 1700 to 1500 cm$^{-1}$ region with less than 5 cm$^{-1}$ are observed in the $N_{py}$ protonated form (bands at 1620 cm$^{-1}$ and 1491 cm$^{-1}$). The same is valid for the o.p. band (B$_1$) 11-$\gamma$CH at about 820 cm$^{-1}$.

The polarized IR-LD spectra of 4-cyanopyridine (4CNPy) are characterized by a significant degree of macroorientation of the molecule's $c$-crystallographic axis. The difference IR-LD spectrum (Figure 4.12) is characterized by positive absorption bands at 1592 cm$^{-1}$ and 1495 cm$^{-1}$ (**8a** and **19a**), with transition moments along the

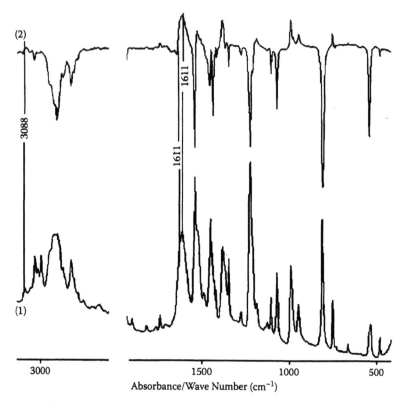

**FIGURE 4.10** (1) Nonpolarized IR and (2) reduced IR-LD spectrum of 4-DMAP after elimination of the **20a** band at 3088 cm⁻¹.

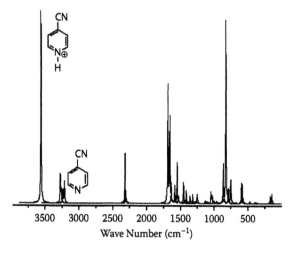

**FIGURE 4.11** Theoretical IR spectra of 4-cyanopyridine and its $N_{py}$ protonated form.

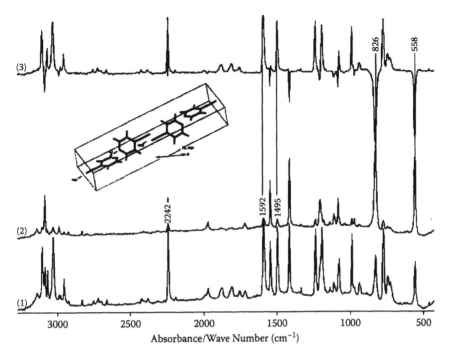

**FIGURE 4.12** (1) Parallel, (2) perpendicular, and (3) difference IR-LD spectra of 4-cyanopyridine (4CNPy).

C-CN band. The positive $v_{CN}$ stretching mode at 2242 cm⁻¹ is expected. All of the molecules in the unit cell of 4-cyanopyridine [259] are oriented in the same way, which resulted in a colinear orientation of the discussed transition moments, such that the elimination of these bands at equal dichroic ratios (Figure 4.12) is to be expected. In contrast, elimination (Figure 4.13) of the o.p. modes at 826 cm⁻¹ (**11-**$\gamma_{CH}$) and 558 cm⁻¹ ($\gamma_{CN}$) leaves second neighboring peaks. Similar to other studied systems, this is a result of the presence in the unit cell of more than one differently oriented molecule. In the case of 4-cyanopyridine, the pyridinium rings of the unit cell molecules closed at angles of 43.3(5)°. The hydrogensquarate salt of the 4-cyanopyridinium cation is characterized by strong overlapping and an underlying Evans's hole effect (Figure 4.14). This state of affairs is typical for the systems with infinite layer structures and strong hydrogen bonding. The highest absorption bands at 3624 cm⁻¹ and 3450 cm⁻¹ correspond to $v_{OH}$ of free and bonded solvent molecules. The broad banding in the whole 3300 to 2080 cm⁻¹ region is attributed to $v_{OH}$ of HSq⁻ fragments participating in strong intermolecular hydrogen bonding. The bands at 1818 cm⁻¹ and 1716 cm⁻¹ belong to $v^s_{C=O}$ and $v^{as}_{C=O}$ of the latter moieties. The o.p. bands, typical for the neutral compound, are observed in the salt as well. This result directly confirms that the aromatic character of the pyridine ring is preserved on $N_{py}$ protonation. Looking at the IR pattern of the hydrogensquarate salt of 4-cyanopyridinium, the strong mutual interaction of the HSq⁻ and $H_2O$ molecules leads to an observation of broad intensive bands, characteristic of their IR spectrum. This is despite the theoretically predicted stronger increase of the intensity of some characteristic

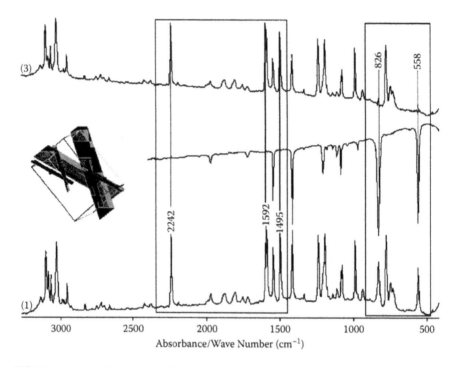

**FIGURE 4.13**    (1) Nonpolarized IR and reduced IR-LD spectra of 4CNPy after elimination of the bands at (2) 1592 cm⁻¹ and (3) 826 cm⁻¹.

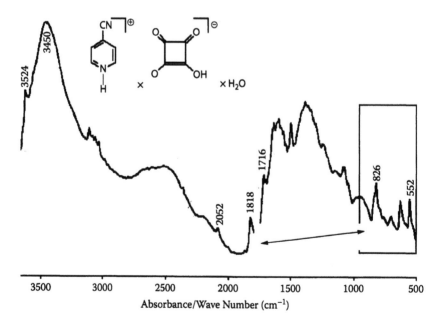

**FIGURE 4.14**    IR spectrum of 4-cyanopyridinium hydrogensquarate monohydrate.

IR bands of $N_{py}$ protonated 4-cyanopyridinium cations. This phenomenon combined with the extremely low frequency shifting of the $v_{OH(Sq)}$ submaximum up to 2080 cm$^{-1}$ suggested that we could describe the system as infinite layers of repeated strong interacting 2HSq$^-$.2H$_2$O fragments, where the cation has a function only to stabilize these layers. This assumption is in accordance with the electronic spectrum of the compound studied, where a significant bathochromic effect of the CT band is observed in the solid state.

In accordance with the data in the literature [258,260,261], the electronic spectrum of 4-aminopyridine shows two characteristic bands of the pyridine aromatic system at 246 nm ($\varepsilon$ = 11056 l.mol$^{-1}$.cm$^{-1}$) and 264$^{sh}$ nm ($\varepsilon$ = 788 l.mol$^{-1}$.cm$^{-1}$). The spectra of protonated compounds contain just one broad band at 255 nm ($\varepsilon$ = 10460 l.mol$^{-1}$.cm$^{-1}$) as also observed for solutions of neutral 4-aminopyridine in 10%, 45%, and 98% H$_2$SO$_4$. In this respect, the UV-spectral study of the 4-aminopyridinium tetrachloropalladate(II) complex salt requires a particular discussion. The UV spectral data of the 4-aminopyridinium tetrachloropalladate(II) complex salt are different. Its UV-spectrum in 0.1 M aqueous solution (equivalent to a solution in 40% H$_2$SO$_4$) exhibits only the broad band at 255 nm (10460 l.mol$^{-1}$.cm$^{-1}$), like the results obtained for solutions in H$_2$SO$_4$ irrespective of the acid concentration. The parameters of this absorption maxima are typical for the imino form and are very similar to those of 1-methyl-1,4-dihydro-4-imino-pyridine and 1-methyl-4-amino-pyridinium salts. As a rule, protonated aminopyridines are obtained as solutions in strong acids (most often H$_2$SO$_4$) and this procedure leads to $N_{py}$ monocations. However, dissolving in concentrated H$_2$SO$_4$ stabilizes the dication structure and the corresponding effects in the aminopyridine UV and NMR spectra have been examined in. Although the relevant equilibria have not been fully clarified, we suggest a 4-aminopyridinium monoprotonation up to 80% H$_2$SO$_4$ with only 98% H$_2$SO$_4$ causing an additional protonation on the exocyclic nitrogen. Our previous study of protonated 2-aminopyridine confirms this conclusion. However, the UV spectra of 2-aminopyridinium sulfate dissolved in 10% H$_2$SO$_4$ are very similar to those of the noncharged 2-aminopyridine, both exhibiting two maxima at about 230 nm ($\pi$-band) and in the 290 to 305 nm range ($\alpha$-band). This result indicates a preservation of the aromatic character of the pyridine ring. In concentrated H$_2$SO$_4$, 2-aminopyridinium sulfate showed only one band at 258 nm ($\varepsilon$ = 5600 l.mol$^{-1}$.cm$^{-1}$) and this effect is not explained by the formation of the imino structure, but by elimination of the n-$\pi$ conjugation and a hypsochromic shift of the $\pi$-spectral band both as a result of the blockage of amino nitrogen.

All the presented data assume a stabilization of the imino form of the protonated 4-AP in the solid state as well as in solution. These data correlate with the known theoretical results, that also indicate the positive charge location in the NH$_2$ group and confirm the experimental stabilization of the imino form after $N_{py}$ protonation. A more detailed comparative structural investigation of protonated aminopyridines and their derivatives should additionally confirm the conclusions thus suggested.

The UV spectra of 3,4-dimethylaminopyridine (3,4-DAP) and its salt (Figure 4.15) show a bathochromic shift of 20 nm for the B-band of the neutral molecule after protonation. This result confirms the assumption of partial charge redistribution and partial distortion of the aromatic character of the salt of 3,4-diaminopyridine. These results are predicted theoretically as well where time-dependence density functional

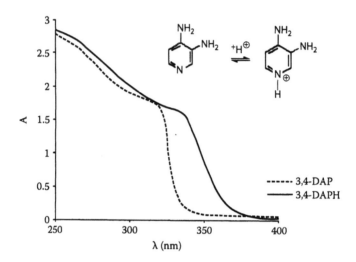

**FIGURE 4.15**   UV spectra of 3,4-DAP (dashed line) and its hydrogentartarate (solid line) in ethanol.

theory (TD-DFT) calculations in ethanol solution give better results, when compared with corresponding configuration interaction single (CIS) data. The protonation of 3,4-diaminopyridine leads to a bathochromic shift of the discussed maximum of 18 nm. The corresponding values are 325 nm (E = 5.20 eV, f = 0.0015) in 3,4-diaminopyridine and 343 nm (E = 4.66 eV, f = 0.2491) in the corresponding $N_1$ protonated form.

In the case of 3,4-diaminopyridine, the bands of its hydrogentartarate and hydro-gensquarate salts were observed at about 328 nm, corresponding to the *B*-band in the neutral form being bathochromically shifted by about 18 and 15 nm, respectively. Comparing with the data of 4-aminopyridine, a large effect of about 5 nm is observed. These results are in accordance with the distortion of the n-$\pi$ conjugation of the 4-$NH_2$ substituent with pyridine ring, which affects the hypsochromic shifting of the discussed maximum. Moreover, our theoretical calculations in ethanol solution give a band at 330 nm with E = 4.72 eV and f = 0.2934.

The electronic spectrum of 4-dimethylaminopyridine shows two characteristic bands for the aromatic pyridine at 255 nm ($\varepsilon$ = 15100 l.mol$^{-1}$.cm$^{-1}$) and 310$^{sh}$ nm ($\varepsilon$ = 654 l.mol$^{-1}$.cm$^{-1}$) [262]. In 10% and 25% $H_2SO_4$ (Figure 4.16) these bands are only slightly affected. However, in 98% $H_2SO_4$ one maximum at 260 nm ($\varepsilon$ = 19200 l.mol$^{-1}$.cm$^{-1}$) and a CT band at 320 nm ($\varepsilon$ = 900 l.mol$^{-1}$.cm$^{-1}$) are observed, suggesting the stabilization of the imino-like form. Similar data were obtained for 4-aminopyridinium tetrachloropalladate(II), but in the 4-aminopyridinium cation the quinolide-like form is still stabilized in 10% $H_2SO_4$. In the case of 4-dimethylaminopyridine, the required higher concentration of the added acid can be explained with the presence of two -$CH_3$ groups in the primary amine. The experimentally obtained data are supported by theoretical calculations, where the neutral form is characterized by a 250 nm band with f = 0.0431, while the pseudo imino tautomer exhibits a band at 260 nm with f = 0.1312. It can be concluded that the system can be described as a pseudo conjugated system of double bonds, as has been predicted theoretically [263].

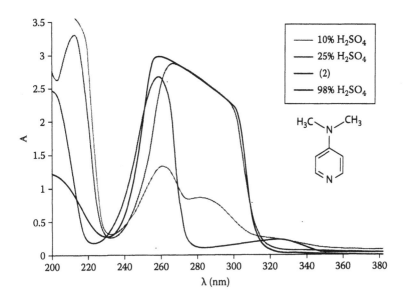

**FIGURE 4.16** UV spectra of 4-dimethylaminopyridine (4-DMAP) in 10%, 25%, and 98% $H_2SO_4$; (2) spectrum of 4-DMAP in $H_2O$.

The UV spectra of 4-cyanopyridine and its hydrogensquarate in $H_2O$ solution (Figure 4.17) are characterized with the bands at about 270 cm ($\varepsilon$ about 12 000 l.mol$^{-1}$.cm$^{-1}$) and 285 cm$^{-1}$ ($\varepsilon$ about 400 l.mol$^{-1}$.cm$^{-1}$), corresponding to the $B$-band of the pyridine fragment and n-$\pi$ transfer of the -CN substituent. In the case of the hydrogensquarate salt in solution, these bands are overlapped with the n-$\pi$ ones of the HSq$^-$ species. These data correlate with our theoretical calculations where the bands at 266 nm (f = 0.7821) and 270 nm (f = 0.9231) are typical for neutral and $N_{py}$ protonated forms of 4-cyanopyridine. In the Vis-region two new bands are observed at 530 nm and 570 nm with $\varepsilon$ values of about 300 and 500 l.mol$^{-1}$.cm$^{-1}$. This characteristic is typical only for 4-cyanopyridinium salts and absent in other different cyanopyridinium salts previously reported. These bands are associated with the red color in solid state of the compounds studied. How can this phenomenon be described? Comparing with the rare studies on the crystallochromy, it is obvious that our system does not correspond to classical ones [227,228]. Looking at the crystal structure, we propose that the red color is a result of a CT band within the layers formed by repeated 2HSq̄2H$_2$O fragments. Additional confirmation follows from the fact that Vis-maxima disappeared after heating the sample at 120°C, at which temperature the included solvent evaporated.

The suggested imino-tautomeric form of the compound studied is also indicated by the $^1$H-NMR spectra. The chemical shift for the pyridine proton signals of 4-aminopyridine belonging to C(3)H and C(2)H are at 6.92 (2H, q) ppm and 8.04 (2H, q) ppm, respectively. The spectra in the solutions of the salts show shifting of the doublet peaks at 7.18 ppm and 8.23 ppm. The data indicate a redistribution of the electron density, a result that can be compared to the analogous data of the 4-aminopyridine/ D$_2$SO$_4$ system, characterized by pairs of doublets at 8.44 ppm and 7.22 ppm. The

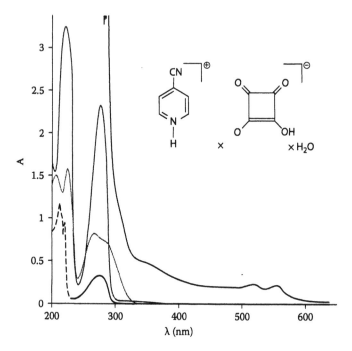

**FIGURE 4.17** UV spectra of 4CNPy (dotted line) and its hydrogensquarate in $2.5 \cdot 10^{-5}$ M water solution; solid-state UV spectra of 4-cyanopyridinium hydrogensquarate monohydrate.

obtained chemical shifts are similar to the previous data of 4-aminopyridine salts and other protonated aminopyridine derivatives.

The chemical shift signals of 3,4-diaminopyridine are observed at 7.83 ppm (H2), 6.95 ppm (H5), and 7.78 ppm (H6), respectively. Comparing with the proton signals of the protonated hydrogentartarate salt, a direct confirmation of the assumption of partial aromatic character distortion in the second case is obtained. A chemical shifting of less than 0.40 ppm and a doublet character of H2 and H5 is observed. These data correlate well with those of 4-aminopyridine, where a difference of 0.3 ppm and a change in the multiplet character of the aromatic proton signals are recorded. The partial charge redistribution in the protonated form of 3,4-diaminopyridine influences the $^{13}$C-NMR signals in a weak manner and a shift of 5 ppm is obtained.

$^{1}$H- and $^{13}$C-NMR spectra of 3,4-diaminopyridinium bis(perchlorate) show chemical shift differences $\Delta\delta$ of the signals, in comparison to the corresponding data of 3,4-diaminopyridine of 0.04 ppm (H2), 0.05 ppm (H5), and 0.13 ppm (H6), respectively, which unambiguously confirms the conclusion that whereas $N_1$ monoprotonation leads to a partial aromatic character distortion, the diprotonated 3,4-diaminopyridine molecule is characterized by full aromatic character. Like in the monoprotonated form, diprotonation practically has no effect on the $^{13}$C-NMR chemical signals and $\Delta\delta$ values of less than 0.1 ppm are obtained.

The chemical shifts for the pyridine proton signals of 4-dimethylaminopyridine, belonging to C(3)H and C(2)H are at 6.88 (2H, q) ppm and 8.98 (2H, q) ppm, respectively.

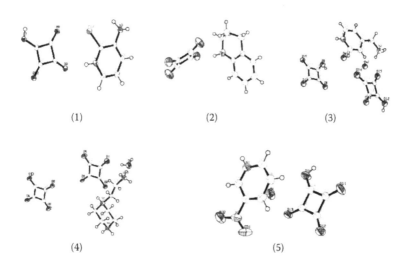

(1)                    (2)                    (3)

(4)                    (5)

**FIGURE 4.18** The molecular structures of (1) 2-chloro-3-aminopyridinium hydrogen-squarate, (2) bis(1,2,3,4-tetrahydroquinolinium) squarate, (3) (4-pyridylmethyl)aminium bis(hydrogensquarate) dihydrate, (4) N-(2-ammoniumethyl)-piperazinium monohydrate hydrogensquarate squarate, and (5) 3-nitropyridinium hydrogensquarate monohydrate showing the atom-labeling scheme. Displacement ellipsoids are drawn at the 50% probability level.

The solution of the protonated salt showed two downfield shifted doublet peaks at 7.00 ppm and 7.86 ppm, thus indicating a redistribution of the electron density.

The $^1$H- and $^{13}$C-NMR spectrum of 4-cyanopyridine exhibited the chemical shift signals at about 7.88 ppm and 7.72 ppm, respectively. Compared to the proton signals of the protonated hydrogensquarate salt, a chemical shift difference of less than 0.10 ppm is obtained. These data are in accordance with theoretically reported ones, where it has been shown that in different cyano-substituted pyridines the $N_{py}$ protonation only weakly affected the aromatic character of the pyridinum ring.

Because of the successful application of polarized IR-LD spectroscopy for analysis of the tautomerism in the different substituted pyridines, this method is a unique experimental tool for assignment of IR bands in the solid state and for detailed analysis of the effects in crystals. For this reason, we have studied a series of hydrogensquarate crystals, where all possible effects of corresponding IR spectra are observed. The following model systems and their explanation will illustrate the applicability of the tool.

The compound 2-chloro-3-aminopyridinium hydrogensquarate crystallizes in the triclinic P-1 space group (Figures 4.18 and 4.19) with Z = 2. The cations and anions form infinite layers by means of the intermolecular interactions Cl···O (2.852 Å), N$^+$H···O (2.613 Å), and HNH···O (2.925 Å). The interlayer distance of 3.338 Å is obtained. The hydrogensquarates form a classical stable α-dimer. The obtained OH···O bond length of 2.544 Å in this structural motif corresponds to the data of other dimers of this type. The reported structure is the second crystallographically determined salt of 2-chloro-3-aminopyridine. The first one was 3-amino-2-chloropyridinium dihydrogenphosphate (CCDC code PIFJAQ [264]). It is interesting to note that in contrast to the last salt in our system, the primary NH$_2$-group participates in asymmetric intermolecular hydrogen bonding, leading to the assumption that the symmetric stretching vibration $v^s_{NH_2}$ will be split in pairs of bands as a result of the FR effect.

Compound bis(1,2,3,4-tetrahydroquinolinium) squarate also crystallizes in the P-1 space group (Figures 4.18 and 4.19), however, the unit cell contains only one cation/

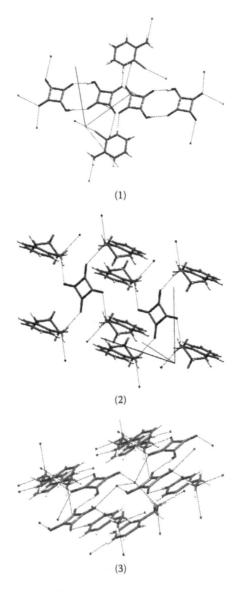

(1)

(2)

(3)

**FIGURE 4.19** Hydrogen bonding in (1) 2-chloro-3-aminopyridinium hydrogensquarate, (2) bis(1,2,3,4-tetrahydroquinolinium) squarate, (3) (4-pyridylmethyl)ammonium bis(hydrogensquarate) dihydrate, (4) N-(2-ammoniummethyl)-piperazinium monohydrate hydrogensquarate squarate, and (5) 3-nitropyridinium hydrogensquarate monohydrate.

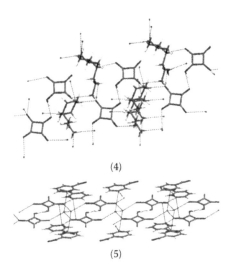

(4)

(5)

**FIGURE 4.19** (Continued) Hydrogen bonding in (1) 2-chloro-3-aminopyridinium hydrogensquarate, (2) bis(1,2,3,4-tetrahydroquinolinium) squarate, (3) (4-pyridylmethyl) ammonium bis(hydrogensquarate) dihydrate, (4) N-(2-ammoniummethyl)-piperazinium monohydrate hydrogensquarate squarate, and (5) 3-nitropyridinium hydrogensquarate monohydrate.

anion pair (Z = 1). This fact allows to us demonstrate the great advantages of the IR-LD spectroscopic tool for IR band assignment in solids and structural elucidation of crystalline compounds. This is also an example of a compound with an isolated squarate dianion as a structural motif. Cations and anions form an infinite chain by an intermolecular $HN^+H\cdots O$ (2.838, 2.743 Å) interaction. The aromatic fragment in the cation is flat, while the torsion angle $N(H_2)\text{-}CH_2CH_2CH_2-$ in the saturated part of the cation is 61.6(3)°. The obtained value is within the range of the other salts of 1,2,3,4-tetrahydroquinoline. In the hydrogen chloranilate salt (PAGXOL [265]) it is 61.6(9)°; in 1,3,5-trinitrobenzene salt ZZZAGJ01 [266] the cation exhibits pronounced disorder and in 3,5-dinitrosalicylate salt GIFNAL [267] the value is 62.0(2)°.

The bis(hydrogensquarate) dihydrate salt of (4-pyridylmethyl)amine was the first diprotonated form of (4-pyridylmethyl)amine to be refined crystallographically. Until now, only a monoprotonated form, stabilized as the hydrochloride (QANWOS [268]) was known. The compound crystallizes in the monoclinic P2/c space group and exhibits Z = 6. The dication, two solvent molecules, and both anions form a 3D network by intermolecular interactions of type $NH_3^+\cdots OH_2$ (2.816 Å), $NH_3^+\cdots O$ (2.807, 2.813 Å), and $N^+H\cdots O$ (2.659 Å). The hydrogensquarates form a new structural motif of isolated α-joined anions ($OH\cdots O$ (2.551 Å) reported for first time in the literature. These pairs of anions are connected to one another by the interactions with solvent water molecules, by means of $HOH\cdots O$ (2.521, 2.800, 2.697, 2.781 Å) as well as $HOH\cdots OH_2$ (2.753 Å) hydrogen bonds (Figures 4.18 and 4.19).

Compound N-(2-ammoniummethyl)-piperazinium monohydrate hydrogensquarate squarate, crystallizes in the noncentrosymmetric monoclinic space group P2₁ with Z = 2 (Figures 4.18 and 4.19). The hydrogensquarate and squarate ions are joined, forming the structural motif, by means of the strong intermolecular interaction of type OH···O (2.511 Å). The cations and anions are connected in a 3D network by means of the additional interactions of type NH···O (2.745, 2.919, 3.027, 2.807, 2.788 Å). Solvent molecules interact with the anions by means of the OH···O hydrogen bonds (2.692, 2.858 Å).

The crystal structure of 3-nitropyridinium hydrogensquarate monohydrate is monoclinic with space group C2/c (Figures 4.18 and 4.19) and Z = 8. The molecules form infinite layers with intermolecular NH···O$_{(HSq-)}$ (2.629 Å) and HOH···O$_{(HSq-)}$ (2.618 and 2.609 Å) hydrogen bonds with participation of cations, anions, and solvent molecules. The formation of stable layered 2HSq$^-$.2H$_2$O fragments is obtained. This structural motif was previously described in 4-cyanopyridinium hydrogensquarate monohydrate (Figure 4.2), and here it is also obtained. It is interesting to mention that like in the case of 4-cyanopyridinium hydrogensquarate monohydrate, the color of the compound in the solid state is red, thus supporting and confirming additionally the phenomenon of the origin of the color in solid state. Hydrogensquarates (HSq$^-$) form stable dimers with equally positioned protons between the anions. The O···O distance between the interacted hydrogensquarates species is 2.461 Å. The molecule of the cation is flat with a maximal deviation from total planarity less than 0.2°.

In the case of 2-chloro-3-aminopyridinium hydrogensquarate, our previous linear-polarized investigation [269] showed bands at 3410 cm$^{-1}$ ($v^{as}_{NH2}$), 3323 cm$^{-1}$/3202 cm$^{-1}$ ($v^{s}_{NH2}$) (Figure 4.20). The second pairs of maxima correspond to FR splitting of the symmetric stretching vibration of the $v^{s}_{NH2}$, usually observed in the system, where the primary amino group participates in asymmetric NH$_2$···X intermolecular interactions. According to the theory described previously, both components will be eliminated at equal dichroic ratios, a fact proven for 2-chloro-3-aminopyridinium hydrogensquarate. This phenomenon was observed and experimentally proven for the first time using the model system 4-aminopyridine [270]. In the last case as well as in the case of 2-chloro-3-aminopyridinium hydrogensquarate, the IR spectroscopic region of the discussed pairs of bands is characterized by broad overlapped absorption bands. For this reason, to prove the expectation that the FR bands of $v^{s}_{NH2}$ are eliminated at equal dichroic ratios, we present here the IR-LD spectroscopic characterization of neutral 2-chloro-3-aminopyridine. The system crystallized in the monoclinic space group P2$_1$/n and Z = 4 (Cambridge Crystallographic Data Centre [CCDC] code YELSEO) [271]. The IR-LD spectra of this compound have well-defined IR maxima and an excellent degree of orientation (as shown in Figure 4.21), using the criteria by Hollas [50], Born and Wolf [51], and Maier and Saupe [52–54]. As can be seen, the elimination of the bands at 3451 cm$^{-1}$ leads to disappearance of the series of benzene in-plane IR bands at 1563 cm$^{-1}$, 1272 cm$^{-1}$, 1141 cm$^{-1}$, and 688 cm$^{-1}$ (Figure 4.21). In all cases the elimination leads to an "inflex point" as a result of the multiple character of these maxima or due to a BS (or DS) effect. This phenomenon has been mentioned previously [272,273]. The multicomponent IR band in this case is split according to the number of molecules in the unit cell into components with different intensities, and these components have different polarizations if the molecules are oriented in a different way. The elimination of the components of the band at 688 cm$^{-1}$ at different dichroic ratios is observed. Like in the case of 2-chloro-3-aminopyridinium hydrogensquarate the IR spectroscopic data for a neutral 2-chloro-3-aminopyridine are in accordance with the crystallographic

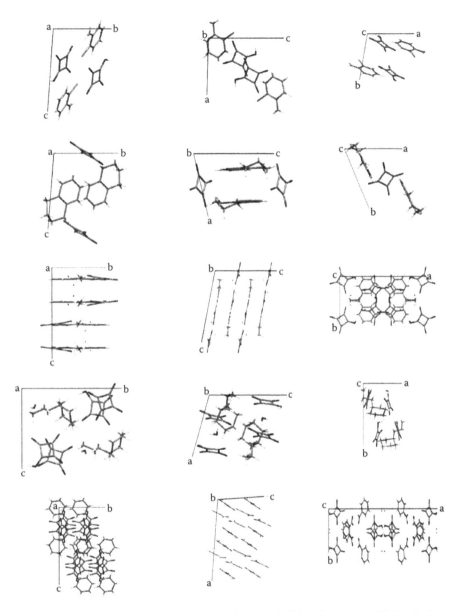

**FIGURE 4.20** View along the *a*, *b*, and *c* axes in the (1) 2-chloro-3-aminopyridinium hydrogensquarate, (2) bis(1,2,3,4-tetrahydroquinolinium) squarate, (3) (4-pyridylmethyl)ammonium bis(hydrogensquarate) dihydrate, (4) N-(2-ammoniummethyl)-piperazinium monohydrate hydrogensquarate squarate, and (5) 3-nitropyridinium hydrogensquarate monohydrate.

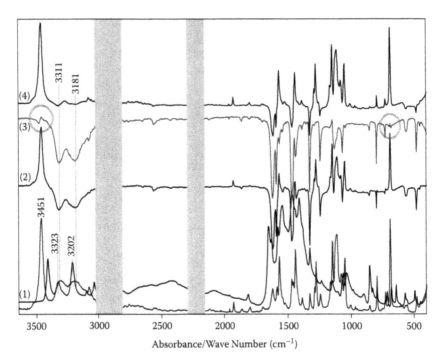

**FIGURE 4.21** Solid-state (1) nonpolarized IR, (2) difference IR-LD, and reduced IR-LD spectra in a nematic host of 2-chloro-3-aminopyridine after elimination of the bands at (3) 3451 cm⁻¹ and (4) 3311 cm⁻¹. IR spectrum of 2-chloro-3-aminopyridinium hydrogen-squarate in nematic mesophase. Gray bars show the self-absorption band of the nematic mesophase. **(See color insert.)**

data. The unit cell of 2-chloro-3-aminopyridine exhibits $Z = 4$ and the molecules are joined into infinite chains by means of a moderate asymmetric intermolecular HNH···N interaction (3.088 Å) [271]. The different molecules are mutually oriented at an angle between the aromatic fragments of 61.9(0)° (Figure 4.21).

The fact that in the solid state, asymmetric hydrogen bonding of HNH···N type is observed leads to an FR splitting of the $v^s_{NH2}$ band into pairs, eliminating at equal dichroic ratios. The bands at 3311 cm⁻¹ and 3181 cm⁻¹ are eliminated simultaneously (Figure 4.21).

In contrast to the corresponding hydrogensquarate salt 2-chloro-3-aminopyridinium hydrogensquarate and the simpler model system 2-chloro-3-aminopyridine, the spectroscopic pattern of bis(1,2,3,4-tetrahydroquinolinium) squarate is characterized by a broad absorption band within the whole 3200 to 2000 cm⁻¹ region (Figure 4.22) belonging to the asymmetric and symmetric stretching vibrations ($v^{as}_{N+H2}$, $v^s_{N+H2}$) of the protonated secondary amine. The broad character of the bands, typical for amino acids and peptides, can be attributed to the FR effect. The corresponding submaxima are eliminated at equal dichroic ratios in the difference IR-LD spectrum. The same effect has been proposed for the carboxylic acid dimers. Due to the strong band overlapping effect observed in this compound, spectral analysis is difficult at a significant level. The interpretation of the IR bands in the whole 4000 to 850 cm⁻¹ region is

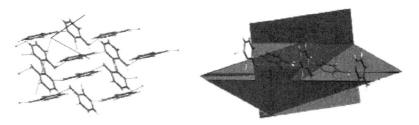

**FIGURE 4.22** Unit cell and hydrogen bonding type in the crystal structure of 2-chloro-3-aminopyridine. (Saha, B., A. Nangia, and J. Nicoud, 2006, *Cryst. Growth Des*, 6:1278–1281. With permission.)

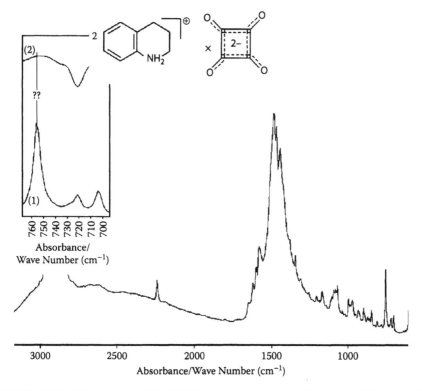

**FIGURE 4.23** IR spectrum of bis(1,2,3,4-tetrahydroquinolinium) squarate (**2**) in nematic host. The inset image shows the (1) nonpolarized IR and (2) reduced IR-LD of the same compound after elimination of the band at 755 cm⁻¹.

not possible and only the bands within the 850 to 400 cm⁻¹ range can be assigned by the reducing-difference procedure (RDP). This example is useful for demonstrating the application of the latter approach for interpretation of the IR spectra of crystalline samples as far as the unit cell of bis(1,2,3,4-tetrahydroquinolinium) squarate contains only one cation/axion pair (Z = 1). The elimination of the band at 755 cm⁻¹ leads to its total disappearance (Figure 4.23).

Similar to 2-chloro-3-aminopyridinium hydrogensquarate, the IR spectra of (4-pyridylmethyl)ammonium bis(hydrogensquarate) dihydrate, N-(2-ammonium-methyl)-piperazinium monohydrate hydrogensquarate squarate, and 3-nitrop-yridinium hydrogensquarate monohydrate, (Figure 4.24) have broad absorption bands between 3100 and 2000 cm$^{-1}$ corresponding to $v_{OH}$ stretching vibrations of the OH-group in the hydrogensquarate anions overlapped with the corresponding stretching modes of the different protonated amino groups. These broad bands have a series of submaxima with a different polarization (obtained by their difference IR-LD spectra) suggesting an FD effect. The bands at about 3500 cm$^{-1}$ belong to the solvent water molecules. The discussed broad band is a typical spectroscopic curve for the hydrogensquarates leading to a difficult interpretation of other IR bands within this region. In some hydrogensquarates the lowest frequency submaxima are obtained about 1900 cm$^{-1}$. More characteristic for hydrogensquarates is the band at about 1800 cm$^{-1}$ belonging to symmetric stretching ($v^s_{CO(HSq)}$) vibrations of HSq$^-$. This maxi-mum can be used to study the DS effect depending on the number of molecules in the unit cell and their mutual orientation. For compounds (4-pyridylmethyl)ammonium bis(hydrogensquarate) dihydrate and N-(2-ammoniummethyl)-piperazinium mono-hydrate hydrogensquarate squarate the elimination of this band leads to an inflex point picture, while in 3-nitropyridinium hydrogensquarate monohydrate total van-ishing of the band is observed, like the effect in bis(1,2,3,4-tetrahydroquinolinium)

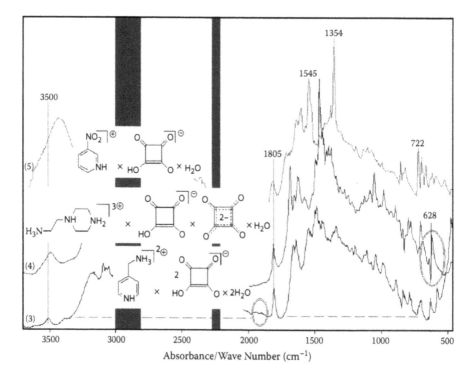

**FIGURE 4.24**   IR spectra of compounds (**3**) through (**5**) (see Figure 4.20) in the solid state in an NLC host.

squarate, facts that are in accordance with the crystallographic data for the orientation of the molecules in the unit cells of IR spectra of (4-pyridylmethyl)ammonium bis(hydrogensquarate) dihydrate, N-(2-ammoniummethyl)-piperazinium monohydrate hydrogensquarate squarate, and 3-nitropyridinium hydrogensquarate monohydrate. The hydrogensquarates have a broad absorption band within the 1700 to 800 cm$^{-1}$ region also as a result of the FD effect. These bands are usually intensive and only in the case of (4-pyridylmethyl)aminium bis(hydrogensquarate) dihydrate, where the cation contains the $NO_2$-group are stronger IR bands belonging to $v^{as}_{NO2}$ and $v^{s}_{NO2}$ at 1545 cm$^{-1}$ and 1354 cm$^{-1}$ are observed (Figure 4.24). The IR characteristics of these compounds are assigned using the region between 850 to 400 cm$^{-1}$, with care being made to take possible EH effects into account.

For the discussed regions, possibilities for interpretation of the IR spectra and the corresponding limitations are common for all of the hydrogensquarates, independent of the structural motifs of these compounds. In the course of our systematic study it could be compared to the corresponding characteristics of those for example of the hydrogensquarate salt of 1,10-phenanthroline, cyclohexylamonium hydrogensquarate monohydrate, with its α-chain, L-argininium and L-serinium hydrogensquarates with the structural motifs, $H_2Sq \cdot HS\bar{q} \cdot 2H_2O$ [227], the nonplanar tetrameric motif of guanidinium hydrogensquarate, the tetrameric motif or isolated hydrogensquarate anion in L-lysinium salts as well as the layered motif $2HS\bar{q} \cdot 2H_2O$, leading to the red color of the 4-cyanopyridinium salt in solid state [274–279].

The analysis of the corresponding ester amides of squaric acid is relatively easier than for hydrogensquarates. For example, 5-amino-2-methoxypyridine ester amide of squaric acid ethyl ester is a compound with proven second-order nonlinear optical (NLO) application in solution and in the bulk. Its structure and properties have been elucidated in detail spectroscopically, thermally, and structurally, using single crystal X-ray diffraction, linear-polarized solid-state IR spectroscopy, UV spectroscopy, and thermogravimetric analysis (TGA), differential scanning calorimetry (DSC), differential thermal analysis (DTA), mass spectrometry (MS), and second-harmonic generation (SHG) methods. Quantum chemical calculations were used to obtain the electronic structure, vibrational data, and NLO properties. At room temperature the compound crystallizes in the noncentrosymmetric space group $C_c$ and exhibits a pseudolayer structure (solid phase 1) with molecules linked by NH···O=C intermolecular hydrogen bonds with a length of 2.955 Å and an NH···O angle of 153.41°, respectively. At 200°C a phase transition is observed, with the solid phase 2 (Figure 4.25) exhibiting new intermolecular NH···N interactions, as elucidated by IR spectroscopy and thermal analysis. The obtained large powder SHG efficiency of 609 times urea confirms the NLO application of the studied compound in the bulk [159,279].

Squaric acid derivatives display a marked proclivity to crystallize in noncentrosymmetric space groups, meaning that such compounds can exhibit NLO properties in the bulk. 5-Amino-2-methoxypyridine ester amide of squaric acid ethyl ester crystallizes in the noncentrosymmetric space group $C_c$ without the help of additional chiral agents. This form we will describe as solid phase 1. Its electronic nonlinearity is based on molecular units containing a strongly delocalized π-electron system with donor and acceptor groups at opposite ends of the molecule, which could explain the SHG results [280,281]. The molecule of 5-amino-2-methoxypyridine ester amide of

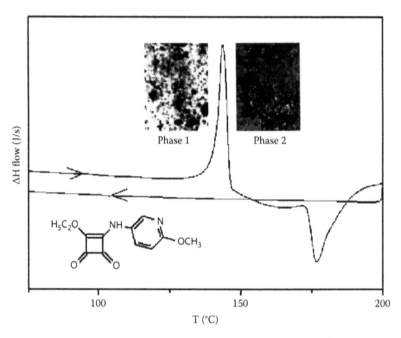

**FIGURE 4.25**  DSC data of 5-amino-2-methoxypyridine ester amide of squaric acid ethyl ester.

squaric acid ethyl ester belongs to the class of two-dimensional quadrupoles such as 4,6-dinitroresorcinol and its derivatives. It is an NLO quadrupole and contains an $N_{(pyridine)}$ atom, which offers a potential center for conversion to a corresponding pyridinium-betaine derivative. The molecules of 5-amino-2-methoxypyridine ester amide of squaric acid ethyl ester are joined into infinite parallel pseudolayers by intermolecular $NH\cdots O=C_{(Sq)}$ hydrogen bonds of length 2.955 Å and an $NH\cdots O$ angle of 153.41°. With the exception of the $OCH_2CH_3$ group, the molecule is virtually flat with a torsion angle between the pyridine and squarate moieties of only 10.7(8)°. The largest deviations from planarity within the pyridine and squarate rings are less than 1.9° and 1.0°, respectively. Similar to Silva et al. [282] the cyclobutene rings are effectively involved in a π-stacking interaction where the interplanar and centroid–centroid distances are 3.844 and 3.554 Å, respectively. The pseudolayer structure of the compound should enable the inclusion of small molecules between the layers, which could then be used for catalysis.

Our quantum chemical calculations confirm the structure of 5-amino-2-methoxypyridine ester amide of squaric acid ethyl ester and a comparison between the predicted and observed bond lengths and angles reveals a satisfactory correlation with differences of less than 0.078 Å and 7.6(7)°, respectively. Two distinct stable conformers of 5-amino-2-methoxypyridine ester amide of squaric acid ethyl ester exist, differing by internal rotation around the C3-O4 bond, where C6-O4-C3-N5 is equal to 0.0° and 178.1°, respectively. When energies are corrected for zero point vibrational energy contributions, the first conformer is more stable by 6 kJ/mol and is characterized by the flat geometry and a maximal deviation from total planarity of 10.3°. The *ab initio* (MP2/6-311++G**)

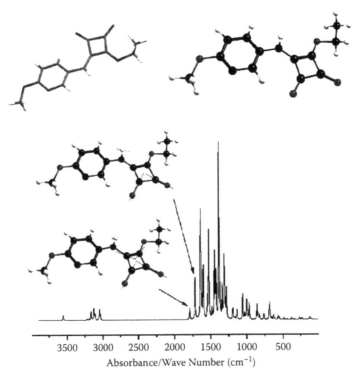

**FIGURE 4.26**  Theoretical predicted geometry of 5-amino-2-methoxypyridine ester amide of squaric acid ethyl ester and the calculated IR spectrum.

approach gives better structural parameters than the DFT (B3LYP/6-311++G**) ones with a difference of 6%. The calculated IR spectrum and some transition moments, important for experimental IR-LD analysis are shown in Figure 4.26.

The evaluated dipole moment of 6.93 Debye, average polarizability ($\alpha = 77.98$ a.u.), and first and second hyperpolarizability ($\gamma = 9964$ a.u.) are all significantly larger than for simple squaric acid [283]. This result is of critical importance when optimizing such materials for an NLO application. For the purpose of comparison with typical push–pull systems, the investigated compound exhibits $\beta_{vec}$ and $\gamma$ values, which are about 123% larger than the corresponding values for p-nitroaniline [227].

The observed significant degree of macroorientation in the polarized IR spectrum of the polycrystalline solid phase 1 facilitates an adequate interpretation of the polarized data and results from the presence of a pseudolayer structure, which adopts a macroorientation in the solid phase toward the orientation director **n** of the liquid crystal. The detailed IR-LD spectroscopic analysis is supported by the theoretical vibrational analysis at B3LYP/6-311++G**.

The obtained macroorientation (Figure 4.27) of the polycrystalline sample determined according to the given quantitative evaluation enables an adequate interpretation of the polarized data and provides the stereostructural information and IR characteristic band assignment. The observed opposite orientations in the difference IR-LD spectrum of the bands at 1798 cm$^{-1}$ and 1694 cm$^{-1}$ confirm their origin

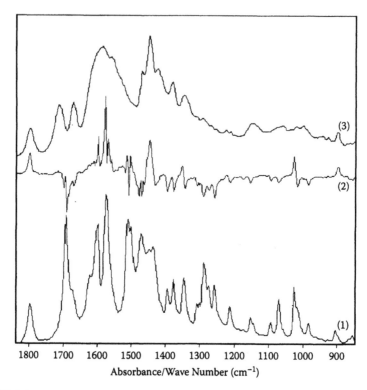

1800   1700   1600   1500   1400   1300   1200   1100   1000   900

Absorbance/Wave Number (cm$^{-1}$)

**FIGURE 4.27**   Solid-state (1) nonpolarized IR and (2) difference IR-LD spectra of solid phase 1; (3) IR spectrum of the solid phase 2 of 5-amino-2-methoxypyridine ester anide of squaric acid ethyl ester.

as $V^s_{C=O(Sq)}$ and $V^{as}_{C=O(Sq)}$ modes. The elimination of the band at 1798 cm$^{-1}$ leads to reduction of the i.p. maximum at 1608 cm$^{-1}$ (calc. 1609 cm$^{-1}$) of the pyridine fragment due to the colinearity of corresponding transition moments. The elimination of the 830 cm$^{-1}$ maximum leads to vanishing of the bands at 746 cm$^{-1}$, 722 cm$^{-1}$, and the broad maximum between 730 and 710 cm$^{-1}$, thus also confirming the vibrational assignments and the structural conclusions due to a colinearity of all o.p. transition moments obtained theoretically and experimentally. The remaining peak at 777 cm$^{-1}$ belongs to the i.p. mode of the pyridine ring (calc. 766 cm$^{-1}$). In contrast to the Raman spectra [284], the peaks are of relatively high intensity and could be used adequately for the purpose of IR-LD spectroscopy.

The experimental UV-VIS spectra exhibit a negative solvatochromic effect, which is revealed by the hypsochromic shift of the long-wavelength charge transfer (CT) band of $\Delta\lambda = 70$ nm with increasing polarity of the solvent (Figure 4.28).

DSC analysis of solid phase 1 (Figure 4.25) shows that an initial distinct endothermic process followed by two overlapping exothermic calorimetric peaks takes place during heating. The endothermic peak at 132°C (with $T_{max}=143.5°C$) must clearly be associated with substance melting with an enthalpy of 155 J/g. The temperature of melting corresponds to the microscopic observations carried out during sample annealing at a constant heating rate. The next broad exothermic DSC

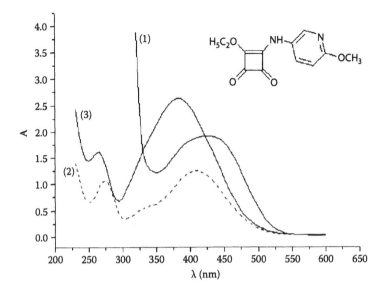

**FIGURE 4.28** UV-Vis spectra of solid phase 1 in: (1) 1,2-dichloroethane ($E_T^N$ = 0.327), (2) CH$_3$CN ($E_T^N$ = 0.460), and (3) H$_2$O ($E_T^N$ = 1.000).

effect, which is partially overlapped with a second exothermic process, is associated with a larger enthalpy change. The second exothermic process ends at 200°C. On combining the results of the DSC analysis with those of the microscopy study, it is important to emphasize that the exothermic DSC effects correspond precisely to the observed formation of a new red solid phase 2. The *in situ* microscopic observation of the solidification reaction(s) supports the DSC result, that is, that the formation and growth of the solid phase starts with a lower rate and then accelerates with increasing temperature. Additional confirmation of the observed solidification reaction is provided by the DSC scan under cooling conditions. Cooling the sample from 200°C to room temperature is not associated with any thermal effects.

To determine the nature of the high temperature solid phase 2, IR spectroscopy was employed as well. The results indicate the formation of new intermolecular interactions. The observed high frequency shift of 8 cm$^{-1}$ for $v^{as}_{CO}$ in the squaric acid anion, and the opposite shift for $\gamma_{C=O}$ of 33 cm$^{-1}$ suggest that these are not affected by intermolecular interactions of the fragment. In contrast, significant low-frequency shifts of the $v_{NH}$ peaks (3100 cm$^{-1}$) and changes for the in-plane (about 1600 cm$^{-1}$) and out-of-plane (850 to 700 cm$^{-1}$) modes of the pyridine ring, that are typical for a symmetry transition from C$_s$ to C$_{2v}$, indicate the formation of strong NH···N$_{(pyridine)}$ intermolecular bonds in the new solid phase 2 (Figure 4.27).

It is interesting to note that a small thermal process has also been detected at 94°C. The shape of this effect indicates a second-order phase transformation, which can be explained by the liberation of a rotation around the C-N bond. This transformation is associated with a small change in the specific heat. It can be concluded that this effect may sterically favor the formation of the new type of intermolecular hydrogen bonds NH···N$_{(pyridine)}$ in solid phase 2.

An additional DTA/TG analysis was performed to provide further support for this explanation of the thermal behavior. The mass losses associated with the endothermic and exothermic reactions detected are indeed very small and cannot be attributed to a release of gaseous product(s) during heating. This result confirms the conclusion that the observed reactions are first melting, and then solidification of a second high temperature phase (solid phase 2). Similar to 4-dimethylamino-N-methyl-4-stilbazolium tosylate (DAST), which crystallizes in the noncentrosymmetric space group $C_c$ [285], solid phase 1 displays a large SHG activity of 609 times that of a urea reference with a 1907 nm fundamental.

The method of the linear polarized IR-LD spectroscopy of oriented colloids in a nematic host could be successfully applied to heterocyclic compounds for experimental assignment of the IR spectroscopic characteristics by means of a correlation with crystallographic data. The observation of the large number of IR bands for these systems leads to a precise spectroscopic and structural assignment using a series of in-plane and out-of-plane vibrations. In addition, the preliminary theoretical vibrational analysis and assignment of the corresponding vibrations to the molecular motions leads practically to characterization of the IR bands in the whole middle IR spectroscopic region using RDP. Effects such as tautomerism, protonation, and specific solid-state interactions can be widely investigated by this tool. The different polarization of the components in the IR spectra of crystals, and in resulting FR and/or FD splitting, BS and EH effects give direct evidence about the origin of submaxima. In the case of a first effect, the components in the multiple band are eliminated at equal dichroic ratios, while in the case of a second effect at a different dichroic ratio. When the molecules in the crystals are oriented in different ways, the bands of a given symmetry class exhibit multiple character, with the number of components equal to the number of the molecules. These components have different dichroic ratios. In contrast, in the cases of parallel-oriented molecules in the unit cell, the BS subbands are eliminated at equal dichroic ratios. For polycrystalline samples, poor quality crystals, or amorphous samples, the method is unique for experimental vibrational assignment of the IR bands to corresponding molecular vibrations. In the amorphous samples, the structural assignment is carried out by the single-molecular approximation.

## 4.2   SMALL BIOLOGICALLY ACTIVE MOLECULES

Intermolecular hydrogen bonds, characterized by low binding energies of typically 10 to 30 kJ mol$^{-1}$ and allowing the biomolecules to interact with their targets before breaking free, are proposed as a basis of the biological activities of molecules in biochemical processes. Therefore, the *in vivo* biochemical processes involve a combination of intra- and intermolecular hydrogen bonding interactions, the details of which are not always well understood.

Amino acids are the simplest biomolecules, containing intramolecular hydrogen bonds, and they serve as building blocks of more complex peptides and proteins. Accurate structure determination and the identification of intramolecular hydrogen bonding motifs in these systems may provide insight into the interactions in larger systems as well as providing valuable experimental data for testing and refining *ab initio* methods. However, polypeptides in the living cell are joined together by amide linkages and the intramolecular hydrogen bonding networks in peptides are better represented by amino amides than amino acids. Amino acid amide derivatives may therefore be regarded as simple peptide models.

The CS-NLC IR-LD spectroscopic tool has been used for experimental vibrational assignment of the characteristic band of all of the essential amino acids, their amides, and small aliphatic and aromatic peptides. The main advantages of the method consist in the possibility of obtaining local structural information for those systems, which are difficult to crystallize as a result of the presence of a chiral center. The scopes and limitation of the method for these systems have been summarized [224–228]. The choice of the amino acid amides was based on their proven pronounced biological activity. Our spectroscopic study was based on a comparison with the known crystallographic data of series of C-α-amidated amino acids. The importance of the C-α-terminal amide for bioactivity can be illustrated by the "potency ratio," which is defined as the relative potency of a peptide amide divided by the relative potency of the corresponding peptide free acid. For example, neurocinin A possesses a potency ratio >10,000 [286]. At present the biological-structural function of the C-α-amide is not fully understood. The amidation arises from the oxidative cleavage of C-α-terminal glycine-extended prohormones. Since the protonated form of amino acid amides exists in the living cell, such information may be useful in understanding the different biological functions of C-α-amide and C-α-acid peptides. In our previous spectroscopic and structural studies of some ester-amides of amino acid amides with squaric acid ethyl ester, L-methioninamide and prolinamide ester amides of squaric acid have, for example, also been reported [287–292].

Herein we present the complete analysis of the newly synthesized compounds L-tryptophanamide esteramide ester amide of squaric acid diethyl ester and L-leucinamide esteramide ester amide of squaric acid diethyl ester with a view to demonstrating the advantages and some limitations of the method using the elucidation by a comparison between the experimental and theoretical vibrational characteristics.

L-Tryptophanamide esteramide ester amide of squaric acid diethyl ester was synthesized, isolated, and structurally determined by single crystal X-ray diffraction (Figure 4.29) [292]. Three-dimensional infinite helix chains of L-tryptophanamide esteramide ester amide of squaric acid diethyl ester are formed by hydrogen bonds of type $_{(indole)}NH\cdots O=C-NH_2$ (2.866 Å) and $O=C-HNH\cdots O=C(Sq)$ (2.932 and 2.851 Å) with participation of indole NH, amide and squaric acid (Sq) fragments, NH, -C=O-NH$_2$, and O=C(Sq). Like other isolated and structurally characterized ester amides of squaric acid, typical values of bond lengths and angles were established in this case. Within the framework of one molecule, the squaric acid fragment is effectively flat with deviations from polarity at 0.7(0)° and 0.7(2)°, respectively. The NH group makes a torsional angle of 6.5(9)* with the squaric acid plane. Contrastingly, the -C=O-NH$_2$ plane is disposed of an angle of 45.7(3)° toward the last discussed structural fragment. The plane of the Sq fragment is inclined at an angle of 17.0(6)° toward the indole plane. Indole and amide planes are mutually oriented at an angle of 46.9(5)°.

View along the *a*-axis

View along the *b*-axis

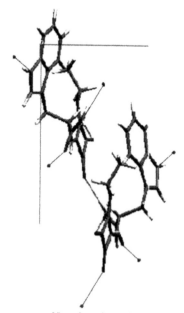

View along the *c*-axis

**FIGURE 4.29** View along *a*, *b*, and *c* axes in L-tryptophanamide esteramide ester amide of squaric acid diethyl ester.

Within the unit cell, the squarate and indole planes of the two molecules of tryptophan amide ester amide of squaric acid are mutually oriented with torsional angles 13.0(3)° and 15.2(0)°, respectively.

Zigzag helix chains are formed through hydrogen bonds of $C=O-NH_2$, the squarate fragment, and NH in L-leucinamide esteramide ester amide of squaric acid diethyl ester (NH$\cdots$O$=C-NH_2$ (2.803 Å), O$=C-NH_2\cdots$O$=C-NH_2$ (2.975 Å), and O$=C-NH_2\cdots$O$=C$(Sq) (3.057 Å)). Like other isolated and structurally determined ester amides of squaric acid diethyl esters, typical values of bond lengths and angles were established in this presented case. Within the framework of an individual molecule, the squaric acid fragment is effectively flat with a deviation from the best plane of only 0.7(1)° (Figure 4.30).

The NH group lies in the squaric acid plane in this second compound, making a torsion angle of 0.3°. Conversely, the $-C=O-NH_2$ plane is disposed at an angle of 88.9(3)° toward the last discussed structural fragment (Figure 4.30). Within the unit cell, the squaric acid planes in the two molecules of leucinamide ester amide of squaric acid ester amides are oriented nearly perpendicular to one another with a torsional angle of 53.1(2)°. Like in the case of L-methionineamide ester amide of squaric acid, the squaric acid fragment is included in the hydrophobic part of the crystal structure.

The conformational analysis of the compound studied in the gas phase was carried out the following way. To generate the $(\phi,\chi)$ potential energy surface, the structures of compound were calculated at the *ab initio* RHF/6–31G*//RHF/6-311++G** level. The geometry is fully relaxed, except for the constrained torsion angles $\phi$ and $\chi$. Values of these angles were chosen by using a step size of 0.5°, within the range from −180° to 180° [293]. The minima observed on the surface were then subjected to full geometry optimization at the B3LYP/6-311++G** level, which should enable correct prediction of the stability order of the calculated minima [294], followed by a second derivative analysis (frequency), which confirmed that all of them were minima. The geometrical parameters of the corresponding energy-minimized conformers were then further investigated. The accessible conformational space of the molecule studied was assumed on the basis of the close resemblance between the Ramachandran contact map and the energy contours map within the limit of 5.0 kcal.mol$^{-1}$ [295], as is also applied elsewhere [296]. The space was calculated using the radial basis function as a girding method. As the overall conformational profiles of modified peptide models can differ from those of common peptide models, the energy-minimized conformers of the investigated molecule are best described by the general shorthand letter notation introduced by Zimmerman [296]. The minimization of energy takes place by optimization of the dihedral angles $\phi$ and $\chi$. The torsion angle $\phi$ is determined by the relative disposition of the cationic NH and electronegative carbonyl groups. The torsion angle $\chi$ is defined as $N(NH)_i-C_j-C_k-C_l$, associated with the conformation in amino acid amide side chains.

Conformational analysis of L-tryptophanamide esteramide ester amide of squaric acid diethyl ester that has been generated by a minimization of the energy in the gas phase takes place by optimization of the dihedral angles. In this case, from the 18 predicted potential energy minima, only six are characterized with $E_{rel}$ lower than 5 kJ/mol (Figure 4.31).

**FIGURE 4.30** Hydrogen bonds and the angle between the amide and squaric acid fragment planes in the molecule L-leucinamide esteramide ester amide of squaric acid diethyl ester.

The predicted geometry parameters (Figure 4.32) correspond to the most stable conformer with $E_{rel}$ equal to 0.0 kJ/mol. The optimized structural parameters as bond lengths and angles agree reasonably well with those obtained by X-ray diffraction. Comparing both the experimental data and the theoretical data presented in Figure 4.32, the atomic distances and angles do not differ by more than 0.087 Å and 8.7(2)°. The listed values agree with the experimental data for other tryptophanamide derivatives studied [297–300]. A significant deviation of some of the dihedral angles of the amide and squarate fragments is, however, determined. The corresponding data as well as the theoretically predicted data suggest an intramolecular hydrogen bond of the type HNH···O=C(Sq) with a bond length of 2.880 Å, which is absent in the solid state. In the condensed state $_{(indole)}$NH···O=C-NH$_2$ (2.866 Å) and O=C-HNH···O=C(Sq) (2.932 and 2.851 Å) hydrogen bond interactions lead to a deviation of the geometry from that of the most stable conformer predicted theoretically. Similar results are obtained concerning di- and tripeptide systems, where the presence of anions or solvent molecules in the unit cell suitable for hydrogen bonding leads to a deviation of the observed geometries from those of the most stable predicted conformers.

Conformational analysis of L-leucinamide esteramide ester amide of squaric acid diethyl ester showed 22 potential energy minima of which only 9 are characterized by $E_{rel}$ lower than 5 kJ/mol. The optimized structural parameters as bond lengths

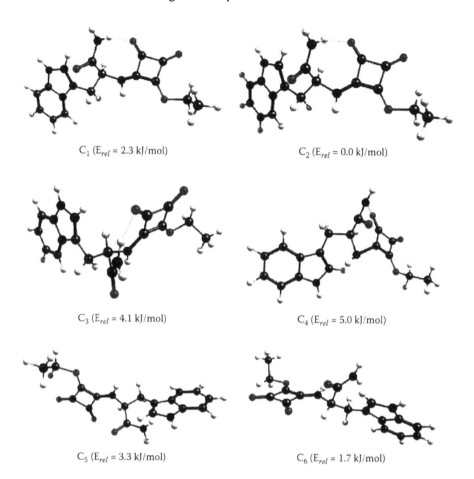

$C_1$ ($E_{rel}$ = 2.3 kJ/mol)      $C_2$ ($E_{rel}$ = 0.0 kJ/mol)

$C_3$ ($E_{rel}$ = 4.1 kJ/mol)      $C_4$ ($E_{rel}$ = 5.0 kJ/mol)

$C_5$ ($E_{rel}$ = 3.3 kJ/mol)      $C_6$ ($E_{rel}$ = 1.7 kJ/mol)

**FIGURE 4.31** Conformers of L-tryptophanamide esteramide ester amide of squaric acid diethyl ester with E less than 5 kJ/mol (most stable one is $C_2$).

and angles agree reasonably well with those obtained by X-ray diffraction. Comparing both the experimental data and the theoretical data presented here, the atomic distances and angles do not differ by more than 0.098 Å, 3.7(5)° and 0.088 Å, 4.8(3)° for the hydrogensquarate and ester amide of squaric acid ethyl ester. The obtained data also correlated well with the experimental data for other leucinamide derivatives, such as DL-acetyl leucine N-methylamide [301], N-acetyl-L-prolyl-L-leucinamide [302], and phenylalanyl-leucinamide hydrochloride monohydrate [339], where the corresponding bond length and angles differences are less than 0.1231 Å, 7.6(6)°.

A significant deviation is apparent on comparing the corresponding data about the dihedral angles in the gas phase and in the solid state. The D(7,5,6,8) and (7,5,4,2) dihedral angles in L-leucinamide esteramide ester amide of squaric acid diethyl ester exhibit deviations of 54.3° and 49.7°, on comparing the experimental and theoretical data. These data could be explained by the presence of strong intermolecular interactions in the solid state leading to the theoretically predicted dihedral angles in the gas phase and those of the isolated molecule. Similar results are obtained for

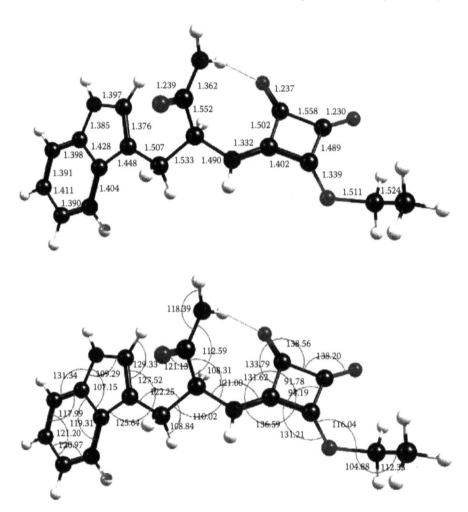

**FIGURE 4.32** Calculated geometry parameters of L-tryptophanamide esteramide ester amide of squaric acid diethyl ester.

small peptides. The amide C=O-NH$_2$ fragment is flat with maximal deviation from a planarity of 0.9°. The squaric acid fragment is likewise effectively planar with a maximal deviation from the plane of 1.3° and is coplanarly disposed toward the NH function with a deviation of the latter group at a dihedral angle of 2.0°. The experimental values are 3.4° and 4.5°, respectively. The planes of the squaric acid moiety and the amide group are cross-oriented at an angle of 71.1° (MP2/6-311++G**) and 66.9°, respectively.

The complicated IR spectral pattern in the NH- and C=O stretching regions of L-tryptophanamide esteramide ester anide of squaric acid diethyl ester (Figure 4.33) requires the determination of the numbers of peaks and their positions by applying the preliminary deconvolution and curve-fitting procedure. The assignment of the data with conventional techniques is difficult, and the experimental confirmation of IR

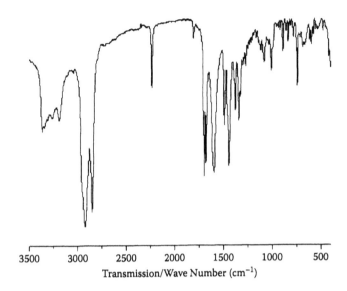

3500        3000        2500        2000        1500        1000        500

Transmission/Wave Number (cm⁻¹)

**FIGURE 4.33**   Solid-state IR spectrum of L-tryptophanamide esteramide ester amide of squaric acid diethyl ester.

characteristic bands as well as the elucidation of structural information have been achieved by the following IR-LD spectral analysis of the compound in solid state, as supported by the crystallographic data presented here. According to the latter data within the framework of the unit cell, a near colinear orientation of the following transition moments is to be expected: (a) $v_{NH}$ and $v_{C=O}$ (Amide I); (b) the out-of-plane frequencies of the Sq ($\gamma_{C-O(Sq)}$) and indole moieties owing to the fact that the planes of these fragments are inclined at a small angle of 17.0(6)°; and (c) Amide I and $\delta_{NH(indole)}$. In the framework of the unit cell, both molecules are oriented so as to render a near colinearity of the $v^s_{NH2}$ and $v^{as}_{NH2}$ stretching frequencies as well as those of corresponding out-of-plane $\gamma_{C-O(Sq)}$ and indole maxima belonging to different molecules.

The application of RDP leads to the following results:

1.  The elimination of peaks at 3347 cm⁻¹ and 1680 cm⁻¹ in equal dichroic ratios indicates their belonging to $v_{NH}$ and Amide I modes. In parallel, a peak at 607 cm⁻¹ is also eliminated, which could be assigned to $\delta_{NH}$(indole) due to the colinearity of its transition moment and Amide I.
2.  The simultaneous disappearance of the bands at 3314 cm⁻¹ and 3195 cm⁻¹ showing that the maxima correspond to $v^{as}_{NH2}$ and $v^s_{NH2}$ modes. In parallel the reduction provokes an elimination of the peak at 1625 cm⁻¹, which is assigned to $\delta_{NH2}$ (Amide II) due to the colinear orientation of $v^s_{NH2}$ and $\delta_{NH2}$ transition moments in the plane of the NH₂ group. The wave number value is of the latter vibration typical for the Amide II mode of primary amides, usually observed in the range of 1615 ± 25 cm⁻¹.
3.  To the indole in-plane modes in the 1800 to 1500 cm⁻¹ region could be assigned the peaks at 1615 cm⁻¹ and 1579 cm⁻¹, which are eliminated at the same dichroic ratio.

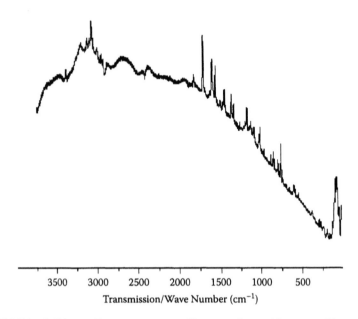

**FIGURE 4.34** Solid-state Raman spectrum of L-tryptophanamide esteramide ester amide of squaric acid diethyl ester.

4. The elimination of the peak at 751 cm⁻¹ leads to the disappearance of those at 462 cm⁻¹ and 426 cm⁻¹ corresponding to out-of-plane indole ring bands and $\gamma_{C-O(Sq)}$ bending vibrations. Other observed bands in the 1850 to 1550 cm⁻¹ range at 1813 cm⁻¹, 1698 cm⁻¹, and 1597 cm⁻¹ belong to $\nu_{C-O(Sq)} + \nu_{C-C(Sq)} + \nu_{C=C(Sq)}$ frequencies. The obtained doublet character of these maxima could be explained with the crystal field splitting effect as far as the pairs of bands are eliminated simultaneously in polarized IR-LD spectra.

The data are typical for other amino-acid amides. The peak at 3370 cm⁻¹ in the NH-stretching region of the compound studied corresponds to the indole $\nu_{NH(r)}$ stretching mode and the 3268 cm⁻¹ peak to a combination mode. All the obtained characteristic peaks for indole ring correlated well with the known ones for the pure amino acid tryptophan, and with theoretical vibrational analysis and IR spectroscopic data of some homo- and heterodipeptides with tryptophyl-fragments. The Raman spectrum of L-tryptophanamide esteramide ester amide of squaric acid diethyl ester is depicted in Figure 4.34.

In IR-spectrum of L-leucinamide esteramide ester amide of squaric acid diethyl ester (Figure 4.35), elimination of the peak at 1583 cm⁻¹ provokes a disappearance of a 3122 cm⁻¹ maximum, thereby establishing their character as $\delta_{NH2}$ and $\nu^s_{NH2}$, as far as within an individual NH₂ group, the corresponding transition moments are colinearly oriented. The simultaneous elimination of $\gamma_{NH}$ and $\gamma_{C=O(Sq)}$ with the Amide I peak in the compound is explained by the fact that within the framework of the unit cell, the neighboring molecules of L-leucinamide esteramide ester amide of squaric acid diethyl ester are mutually oriented at an angle of 53.1(6)°. This results

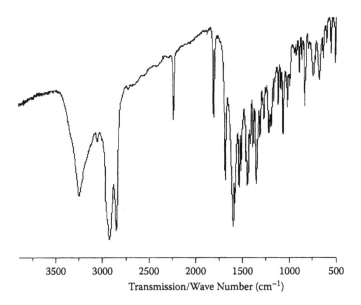

**FIGURE 4.35** Solid-state IR spectrum of L-leucinamide esteramide ester amide of squaric acid diethyl ester.

in a near to colinear orientation of the out-of-plane modes $\gamma_{NH}$, $\gamma_{C=O(Sq)}$, whose transition moments are disposed perpendicularly to the plane of the squaric acid residue and the $\nu_{C=O}$ (Amide I) modes. The Raman spectrum of the compound is depicted in Figure 4.36.

The $^1$H-NMR spectrum of TrpA shows $\alpha_{CH}$, $\beta_{CH2}$, and $\beta_{CH2}$ proton signals of tryptophyl side chains at 4.31 ppm, 3.39 ppm, and 3.55 ppm, respectively. The chemical shifts of aromatic protons are observed in the 7.30 to 7.80 ppm range, including the NH signals at 7.60 ppm. These data correlated with the values observed for the pure amino acid reported by Inoue at al. [340]. The signals at 1.1 and 1.3 ppm correspond to $CH_3$ and $CH_2$ protons of ethyl ester fragment in the molecule.

$^1$H-NMR spectra of L-leucinamide esteramide ester amide of squaric acid diethyl ester are characterized by chemical shifts of the leucine residues of $\alpha_{CH}$, $\beta_{CH2}$, $\gamma_{CH}$, and $\delta_{CH3}$ at 0.990, 1.719, 3.298, and 3.908 ppm. Two additional signals are observed at 1.807 and 4.992 ppm, corresponding to $CH_3$ and $CH_2$ protons in the ethyl ester group in squaric acid residue.

In this section of the book we will also focus on some other classes of the small biologically active compounds, which could be analyzed successfully with the described CS-NLC orientation tool. Discussion of the limitations of the method for some systems is also included. Serotonin is a neurotransmitter of enormous biological importance. Despite its simple molecular structure, serotonin plays a significant role in a number of fundamental biological processes, including the regulation of mood, stress, sleep, eating cycles [303], and muscle and gastrointestinal function. More recently, it has been shown that serotonin also plays a key role in heart disease and asthma. At physiological pH, serotonin is protonated [304] and not surprisingly there are a variety of receptor sites involved in these diverse circumstances.

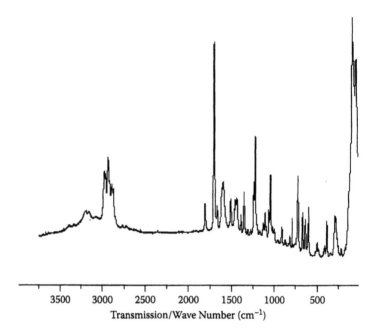

**FIGURE 4.36**  Solid-state Raman spectrum of L-leucinamide esteramide ester amide of squaric acid diethyl ester.

Serotonin's versatility is likely influenced by its ability to reorient the flexible ethylamine side chain and 5-OH groups in matching up with differences in the receptor sites. By studying the conformational preferences inherent to the isolated molecule serotonin, we hope to contribute to a better understanding of how serotonin might interact with these receptor sites to carry out its many biological functions. The inherent conformational preferences of serotonin are also interesting from a fundamental viewpoint as a prototypical flexible biomolecule. A long-term goal of studies of isolated gas-phase serotonin would be to map the potential energy surface for isomerization in as complete a fashion as possible. This would include determining the spectroscopic signatures of the low-energy conformational isomers and their fractional abundances, and measuring the relative energies of the minima and energy barriers separating these minima. In so doing, one could learn about the efficient isomerization pathways on the multidimensional potential energy surface. Serotonin is a close analogue of tryptamine, differing only in its hydroxyl-substituent in the 5-position on the indole ring. Tryptamine is a molecule that itself has played a foundational role as a prototypical flexible biomolecule [304–316]. Tryptamine is a biogenic serotonin-related indoamine and is the decarboxylation product of the amino acid tryptophan. Interest has recently focused on the role of tryptamine in a number of biological systems.

The interest of tyramine is based on the fact that it is a catecholamine-related indirect sympathomimetic amine and occurs widely in plants and animals. Since a G protein-coupled receptor with a high affinity for tyramine, called TA1, has been discovered in brain and peripheral tissue, tyramine is also discussed as a direct

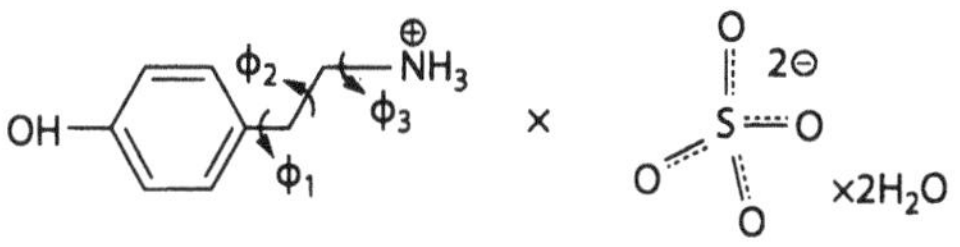

**FIGURE 4.37**   Chemical diagram of bis(tyrammonium) sulfate dihydrate.

neurotransmitter that regulates neuronal activity in basal ganglia. The full under-standing of the receptor-binding manner of small molecules requires knowledge of the conformational preferences of these molecules, which depend mainly on their protonation. Despite this state of affairs, experimental studies of correlations between structural and spectroscopic properties are rare for this class of compounds. Only the crystal structures of several tyrammonium salts have been previously reported [317–322].

The molecular flexibility of these systems leads to an investigation of the con-formational preference of tyramine and tryptamine in the gas phase and in solution [323–330]. Due to the protonation/deprotonation, tautomeric and conformational changes are the most important, nonmetabolic structural transformations through-out the drug delivery process and competitive receptor binding for small, biologi-cally active molecules. In the partitioning phase, the molecules (mostly synthetic drug molecules with oral administration) may have to traverse both hydrophilic and hydrophobic phases. A possible way of reducing the free energy in the actual chemical environment is just the alteration of the protonation state and/or formation/disruption of intramolecular hydrogen bonds by conformational changes. For these reasons we focused on these molecules and studied the conformational preference of their protonated forms in the solid state depending on the type of the counter ion, that is, varying the intermolecular interactions. The interesting results reflect the possibilities and the limits of the IR-LD spectroscopic method.

We present first the model system bis(tyrammonium) sulfate dihydrate [195] (Figure 4.37) crystallizing in the monoclinic space group $P2_1/c$; its structure con-sists of a 3D supramolecular network of tyrammonium cations, water, and sulfate anions directed by intermolecular hydrogen bonding interactions. Two nonequiva-lent tyrammonium cations are observed with pseudo $T$ and $T$ *trans* configurations, respectively (Figure 4.37). The usage of this system demonstrates the applicability of the polarization method to define the IR spectroscopic characteristics of the two nonequivalent molecules in the unit cell.

One tyrammonium cation participates in intermolecular interactions of types $_{(Tyr)}OH\cdots OSO_3^{2-}$ (2.757(2) Å), $_{(Tyr)}NH_3^+\cdots OH_2$ (2.787(3) and 2.840(3) Å), and $_{(Tyr)}NH_3^+\cdots OSO_3^{2-}$ (2.849(3) Å), whereas the second cation is involved in a different pattern of intermolecular hydrogen bonding of $_{(Tyr)}NH_3^+\cdots OH_2$ (2.962(2), 2.900(2), and 2.912(2) Å), $_{(Tyr)}OH\cdots OSO_3^{2-}$ (2.674(2) Å), and $HOH\cdots OH_{(Tyr)}$ (2.961(3) Å). In addition, the anions and solvent molecules interact through moderate hydrogen bonds of the type $HOH\cdots OSO_3^{2-}$ with lengths of 2.876(2) and 2.825(3) Å, respec-tively (Figure 4.38). Differences of 67.37°, 4.5°, and 7.37° are observed for the tor-sion angles $\phi_1$-$\phi_3$ (Figure 4.37), meaning that the cations exhibit $T$ and pseudo $T$ *trans* configurations, respectively. A CCDC database investigation for other tyrammonium

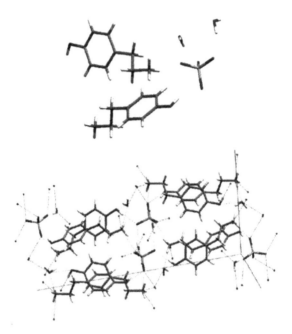

**FIGURE 4.38**  Unit cell and hydrogen bonding of bis(tyrammonium) sulfate dihydrate. (Koleva, B.B., Kolev, T., Zareva, M., Lamshöft, S., and M. Spiteller., 2008, *Trans. Met. Chem.*, 33:911–919. With permission.)

salts (BAKMOP: 3-indole-acetic acid tyramine [319]; TYRAMC: tyramine hydrochloride [320]; TYRTHM: tyramine 1-thyminyl-(acetic acid) complex monohydrate [321]; XUDRIX; bis(tyrammonium) ferrocene-1,1′-dicarboxylate [321]) revealed that effectively ideal *T trans* conformers are typically stabilized for the cation. Only in the case of bis(tyrammonium) ferrocene-1,1′-dicarboxylate [322] with its two nonequivalent tyrammonium cations, were contrasting *T* and *G trans* conformers found in the solid state [329]. In the unit cell, the aromatic benzene rings adopt interplanar angles of 14.72°, 51.70°, 18.84°, 50.03°, 53.67°, and 19.95°.

Our theoretical calculations can be compared with previous detailed studies of the tyramine neutral molecule, its zwitterion, and its hexahydrate at DFT/B3LYP/6-31G*, DFT/B3LYP/6-311++G* [329], and MP2/6-31G* levels [324]. In the present theoretical approximation we investigated the tyraammonium cation as its heptatydrate, with respect to the intermolecular interactions of one cation with five oxygen atoms of solvent water molecules and sulfate anions as taken from the crystallographic data in solid state. The molecular structure in aqueous solution was optimized at the MP2/6-31++G* level of theory and basis set. The resulting most stable conformation of the systems studied is characterized by the torsion angles $\phi_1$, $\phi_2$, and $\phi_3$ (Figure 4.37) of 93.89°, 174.32°, and 178.94°, respectively, meaning that the *T trans* configuration is preferable. These data correlate well with previously reported values and differences of only 6.09°, 3.88°, and 0.76° are obtained on comparing the $\phi_i$ ($i$ = 1-3) values. The calculated pattern of intermolecular hydrogen bonding is in reasonable agreement with the experimental data with the $NH_3^+$ group participating in three $NH_3^+\cdots O$ interactions

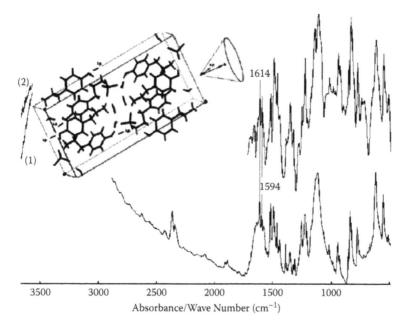

**FIGURE 4.39** (1) Nonpolarized IR and (2) difference IR-LD spectra of bis(tyrammonium) sulfate dehydrate. Orientation is along the *b* axis of the solid bis(tyrammonium) sulfate dihydrate toward the orientation director **n** of the nematic host.

with bond lengths of 2.579, 2.554, and 2.761 Å, respectively. The OH group interacts with water molecules both as a hydrogen donor and as a hydrogen acceptor, and H-bonds $_{(Tyr)}OH\cdots OH_2$ and $HOH\cdots OH_{(Tyr)}$ are predicted with respective lengths of 2.431 and 2.566 Å. The water molecules participate in strong $HOH\cdots OH$ hydrogen bonds with distances of 2.3822, 2.516, 2.710, and 2.658 Å. The comparison with the crystallographic data indicates respective $\phi_i$ differences of 53.96°, 103.28° ($\phi_1$); 0.65°, 4.18° ($\phi_2$); and 0.01°, 7.38° ($\phi_3$) for the nonequivalent cations in the unit cell. Thus the employed theoretical approach predicts the preferred conformation of the tyrammonium cation in a reasonable manner, in particular for the torsion angles $\phi_2$ and $\phi_3$.

The nonpolarized IR and difference IR-LD spectra are depicted in Figure 4.39. The observed significant degree of orientation of suspended particles in the nematic host allows a relatively precise interpretation of polarized IR spectroscopic data. The positive absorption maxima of the in-plane stretching vibrations of the benzene ring with the transition moment oriented along the main symmetry axis and disposed in the plane of the skeleton, that is, **8a** and **19a** (1614 cm⁻¹ and 1594 cm⁻¹) for the in-plane mode of 1,4-disubstituted benzene in the tyrammonium cation, indicate the macroorientation depicted in Figure 4.39. The individual transition moments of the discussed maxima are disposed along the b-axis in the unit cell and the observation of positive bands means that they must adopt approximately colinear orientations toward the orientation director **n** or according to theory must be placed within the range 0° to 54.7° (magic angle).

The interpretation of the IR spectrum of the studied compound is rather difficult within the 3500 to 2500 cm⁻¹ IR region owing to strong overlapping of the bands corresponding to stretching OH modes ($\nu_{OH}$) of the 4-OH substituent in the tyrammonium cation and in water, all of which participate in moderate intermolecular interactions. The 1700 to 1450 cm$^{-1}$ region contains a series of bands between 1680 and 1625 cm$^{-1}$ corresponding to $\delta^{as}_{NH3+}$, $\delta^{as'}_{NH3+}$, and $\delta_{H2O}$ bending vibrations. The bands at about 842 cm$^{-1}$ belong to the out-of-plane mode of 1,4-disubstituted benzene. The most intensive bands in the IR spectrum at 1117 cm$^{-1}$ and 618 cm$^{-1}$ correspond to the stretching and bending modes of the sulfate anion.

Experimental confirmation of the assignment of i.p. and o.p. modes is obtained by their consequent elimination at different dichroic ratios. The elimination of 1614 cm$^{-1}$ leads to the disappearance of the 1594 cm$^{-1}$ band, due to their belonging to the same symmetry class. The elimination of the band at 842 cm$^{-1}$ gives rise to the bands at about ±4 cm$^{-1}$ with the same origin, but belonging to the other differently orientated molecules in the unit cell.

Single-crystal X-ray diffraction data of bis(tyrammonium) sulfate dihydrate show that the structure consists of a 3D network of cations, anions, and solvent molecules linked by intermolecular interactions. The two nonequivalent tyrammonium cations with pseudo $T$ and $T$ $trans$ conformations differ with respect to the types of their intermolecular interactions and by their dihedral angles $\phi_i$ ($i - 1 = 3$). One cation participates in $_{(Tyr)}$OH$\cdots$OSO$_3{}^{2-}$ (2.757(2) Å), $_{(Tyr)}$NH$_3{}^+\cdots$OH$_2$ (2.787(3) and 2.840(3) Å), and $_{(Tyr)}$NH$_3{}^+\cdots$OSO$_3{}^{2-}$ (2.849(3) Å) bonds, and the second one in $_{(Tyr)}$OH$\cdots$OSO$_3{}^{2-}$ (2.674(2) Å), $_{(Tyr)}$NH$_3{}^+\cdots$OSO$_3{}^{2-}$ (2.962(3), 2.900(3) and 2.912(3) Å), and HOH$\cdots$OH$_{(Tyr)}$ (2.961(2) Å) interactions, respectively. The corresponding $\phi_i$ values are (a) pseudo $T$ $trans$—39.95 ($\phi_1$), 173.67 ($\phi_2$), and 178.93 ($\phi_3$); and (b) $T$ $trans$—107.18 ($\phi_1$), 178.17 ($\phi_2$), and 171.56 ($\phi_3$), respectively. Both nonequivalent cations interact with five oxygen atoms. This system is reasonably described by the performed MP2/6-31++G* quantum chemical calculations on the tyrammonium cation heptahydrate. The most stable calculated conformation is characterized by torsion angles $\phi_1$, $\phi_2$, and $\phi_3$ of 93.89°, 174.32°, and 178.94°, respectively, indicating that the $T$ $trans$ configuration is energetically preferable. A comparison with the crystallographic data indicates $\phi_i$ differences of 53.96°, 103.28° ($\phi_1$); 0.65°, 4.18° ($\phi_2$); and 0.01°, 7.38° ($\phi_3$) for the nonequivalent molecules in the unit cell, which means that a reasonable theoretical prediction of $\phi_2$ and $\phi_3$ is achieved. In addition, the following intermolecular hydrogen bonding interactions are predicted: NH$_3{}^+\cdots$O with bond lengths of 2.579, 2.554, and 2.761 Å, $_{(Tyr)}$OH$\cdots$OH$_2$ and HOH$\cdots$OH$_{(Tyr)}$ (2.431 and 2.566 Å), and strong HOH$\cdots$OH hydrogen bond with lengths of 2.3822, 2.516, 2.710, and 2.658 Å, respectively. Experimental IR band assignment of the nonequivalent cations in the solid state was performed by means of linear polarized IR spectroscopy of oriented colloids in a nematic host leading to differences in the corresponding frequencies of ±4 cm$^{-1}$.

Tryptammonium hydrogentartrate [196] (Figure 4.40) is our second example on a biologically active amine. It is crystallized in the orthorhombic space group P2$_1$2$_1$2$_1$ and exhibits a crystal structure consisting of the tryptammonium cation, hydrogentartarate anion, and a solvent methanol (Figure 4.41). The species are joined by intermolecular NH$\cdots$OH(CH$_3$) (2.998 Å) and NH$_3\cdots$O(tart) (2.772, 2.902 and 2.847 Å) hydrogen bonds into a 3D network. Hydrogentartarate ions are themselves connected

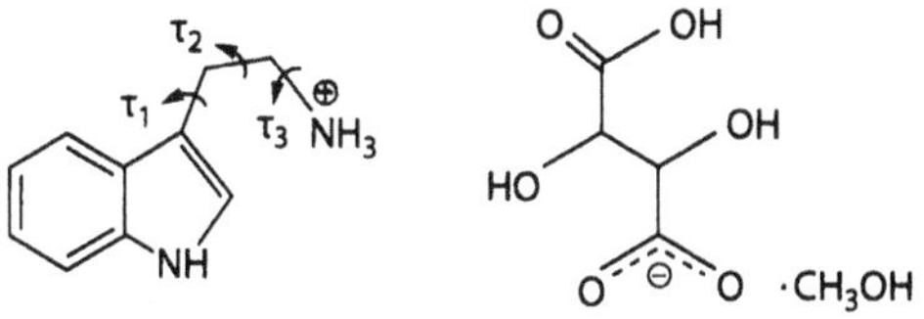

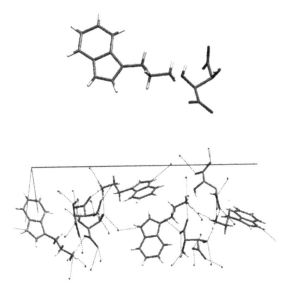

**FIGURE 4.40** Chemical diagram of tryptammonium hydrogentartarate and definition of the torsion angles $\tau_1$ to $\tau_3$.

**FIGURE 4.41** Unit cell and linkage of the molecules of tryptammonium hydrogentartarate into 3D networks. (Koleva, B.B., Kolev, T., Seidel, R.W., Mayer-Figge, H., and Sheldrick, W.S., 2008, *J. Mol. Struct.*, 888:138–144. With permission.)

into infinite chains by strong intermolecular O···O interactions with a bond length of 2.481 Å. They participate in moderate hydrogen bonding as well with the solvent molecule $_{(tart)}$O···O(CH$_3$) 2.736 and 2.762 Å). The indole ring is effectively flat with a maximal deviation from planarity of 0.4°. The geometry parameters correlate well with the values obtained from other crystallographic refinements for the tryptammonium cation [330]. The comparison of the data reveals differences of less than 0.0021 Å and 0.2°, respectively. The obtained torsion angle values $\tau_1$, $\tau_2$, and $\tau_3$ (Figure 4.41) for the tryptammonium cations are 33.6(3)°, 179.0(3)°, and 162.3(3)°, respectively.

The theoretical calculations for the interaction of the tryptammonium cation and two water molecules provide the most stable conformer shown in Figure 4.42. The $E_{rel}$ is 0.4 kJ/mol and the $\tau_1$, $\tau_2$, and $\tau_3$ values are 28.3°, 180.0°, and 176.4°, respectively. As can be seen, the data correlates with the experimental values with differences of $\Delta\tau_1 = 5.3°$, $\Delta\tau_2 = 0.1°$, and $\Delta\tau_3 = 14.1°$, respectively. Other geometrical parameters, that is, bond lengths and angles, differ in comparison with the

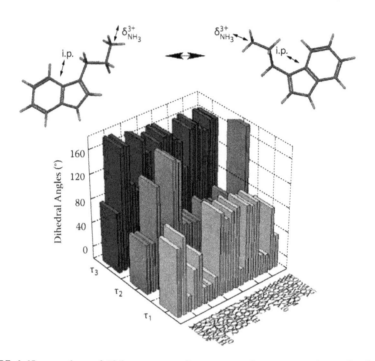

**FIGURE 4.42** $\tau_i$ values of 19 known crystal structures of tryptammonium salts depicted with their CCDC codes. DAMNAH (tryptamine benzoic acid) [332]; DOTXUF10 (7-methylguanosine-5′-monophosphate tryptamine trihydrate) [333]; FINZIM ((P)-tryptamine 4-chlorobenzoic acid) [334]; GANFOQ (tryptamine 3,4-dimethoxybenzoate) [335]; JOZCOQ (tryptammonium biphenyl-3-carboxylate) [336]; JOZMEQ (tryptammonium biphen-4-ylacetate) [336]; LACLOQ (tryptamine 2-thiophenecarboxylic acid); LACQAH (tryptamine 3-indoleacetic acid) [337]; TIFLAV (tryptammonium 3-phenylpropionate) [337]; TRYPIC (tryptamine picrate) [338]; XIJAP (tryptamine trans-3-nitrocinnamic acid) [339]; TRPAC (tryptammonium phenylacetate) [340]; TRYPTA10 (tryptamine hydrochloride) [341]; WODTUE (tryptamine trans-cinnamic acid); WODVAM (tryptamine phenoxyacetic acid); WODVEQ (tryptamine (phenylthio)acetic acid); WODVIU (tryptamine 1-naphthylacetic acid); WODVOA (tryptamine p-nitrocinnamic acid); and WODVUG (tryptamine diphenylacetic acid) [342].

experimental crystallographic data by less than 0.031 Å and 2.3°, respectively. The $NH_3^+$ group in the cation interacts strongly with one of the water molecules with an NH···O bond length of 2.531 Å. The indole ring is flat with a deviation from total planarity of only 0.02°.

A surprising result is obtained when comparing our experimental and theoretical data with previously studied tryptammonium-containing systems with regard to the conformational preference of the cation. Until now, the structures of about 19 tryptammonium salts have been crystallographically refined and the observed torsion angles $\tau_i$ ($_i$ = 1-3, Figure 4.40) are depicted in Figure 4.42. In the majority of the systems, the $\tau_3$ and $\tau_2$ values are correspondingly higher than 150° and lower than 80°, respectively. For $\tau_1$ values two characteristic angle ranges are observed: less

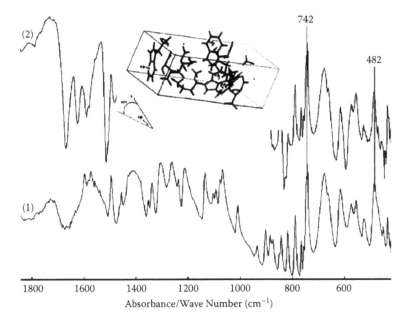

**FIGURE 4.43** (1) Nonpolarized IR and (2) difference IR-LD spectra of tryptammonium hydrogentartarate.

than 35° and between 100° and 120°, with the second range being observed more frequently (Figure 4.42). These results confirm that the theoretical approximation model describes relatively well all the known conformers of the tryptammonium molecule in the solid state as solved by X-ray diffraction. The $\tau_i$ values of the cation are different to those of the neutral tryptamine molecule, where the crystal structure exhibits angles of 89.4(2)°, 60.7(4)°, and 78.0(4)°, respectively [331].

The conformational preference of the tryptammonium cation and its derivatives can be studied successfully using linear-polarized IR spectroscopy of oriented colloids in an NLC suspension. However, the typically observed complicated IR spectral pattern of tryptammonium hydrogentartarate (Figure 4.43), both in the NH- and C=O stretching regions, requires the determination of the numbers of peaks and their positions by applying preliminary deconvolution and curve-fitting procedures. The assignment of the data with conventional techniques is difficult, and experimental confirmation of the characters of the typical IR bands as well as the extraction of structural information could be achieved only by IR-LD spectral analysis in the solid state, through a comparison with crystallographic data. This poses the question as to when it is possible to determine at a significant level, which of the most frequently observed conformers in a solid state is obtained for a particular tryptammonium-containing system.

So, starting with the analysis of the difference IR-LD spectrum of tryptammonium hydrogentartarate in NLC suspensions, a reasonable orientation of solid particles was observed (Figure 4.43). The recorded positive absorption maximum corresponding to the out-of-plane modes of the indole ring at 742 cm⁻¹ and 482 cm⁻¹ indicated an orientation of the molecules toward the NLC director **n** as shown in Figure 4.43. In the context of the unit cell, the summarizing transition moments of the indole ring are oriented along the *b* axis and the positive orientation in the corresponding difference spectrum suggests that the transition moments make an angle with director **n** within the range of 0.0° to 54.7°. Confirmation of the proposed assignment is provided by the simultaneous elimination of the bands at an equal dichroic ratio.

From the spectroscopic point of view within the context of isolated molecules, the i.p. and o.p. modes adopt similar angles with other characteristic IR maxima for asymmetric and symmetric stretching and bending modes of the $NH_3^+$ groups, and this state of affairs could be used for structural elucidation in the solid state. As to be excepted in our system, a subsequent elimination of the bands at 1614 cm$^{-1}$ and 1579 cm$^{-1}$ (i.p. indole modes) with those at 1683 cm$^{-1}$ and 1544 cm$^{-1}$ is observed in accordance with the geometry of the cation, where pairs of these maxima are oriented in a mutually colinear manner. However, this analysis is also valid for the second conformer. The picture is also complicated by the presence of more than one differently oriented molecule in the unit cell of the $P2_12_12_1$ space group (Z = 4). This means that in the case of the tryptammonium cation, a limitation of the IR-LD spectroscopic tool of oriented colloids in NLC suspension is apparent, which prevents a certain assignment to one of the two observed conformers of the molecule in the solid state. However, for the vibrational assignment of the characteristic bands to corresponding vibrational modes, the method is apparently unique as far as allowing the investigation of solids independent of their crystalline or amorphous character.

This part of the book illustrates the successful application on polarized IR spectroscopy of oriented colloids in a nematic host for vibrational assignment and local structural elucidation of small biologically active compounds. However, we have tried to show the readers that for systems possessing hydrogen bonding functional groups leading to possible intra- or intermolecular interactions, like, for example, amino acids and their amides, the comparison between the theory and experiment must be performed carefully and in every case supported by the other individual experimental methods, such as Raman spectroscopy, nuclear magnetic resonance, or the ideal case of single-crystal XRD. In some of these systems, modern quantum chemical approaches usually give the most stable conformer geometries with predominant intramolecular interactions, which affect the predicted vibrational characteristics. In contrast, the crystal structure determinations in our investigations show that these molecules participate predominantly in intermolecular interactions, which makes the interpretation of the polarized IR-LD spectra more difficult. One of the most successful approaches is when the system is characterized crystalographically so as to provide the experimental geometry as an input file for the quantum chemical calculations.

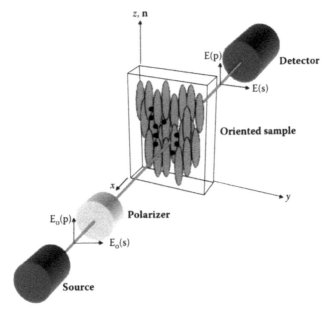

**FIGURE 1.1** Optical scheme for a measurement of IR-LD spectra: source of irradiation, polarizer, oriented sample, and detector.

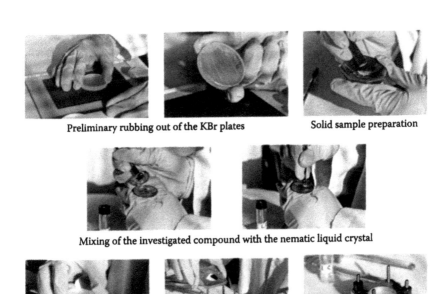

Preliminary rubbing out of the KBr plates        Solid sample preparation

Mixing of the investigated compound with the nematic liquid crystal

Placement of the suspension between the KBr plates and then in the holder

**FIGURE 2.2** Preparation of the oriented colloidal suspension in a nematic liquid crystal.

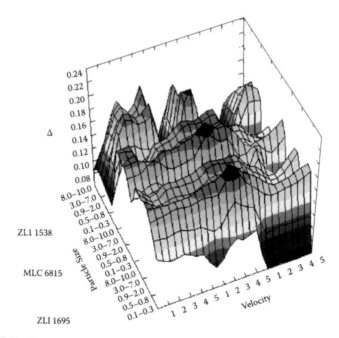

**FIGURE 2.7** 3D graph with the dependence of the parameter Δ versus particle size (μm), the velocity of the slippage (μm/s), and the layer thickness of the suspension of DL-isoleucine in the nematic liquid crystals ZLI-1695, MLC-6815, and ZLI-1538.

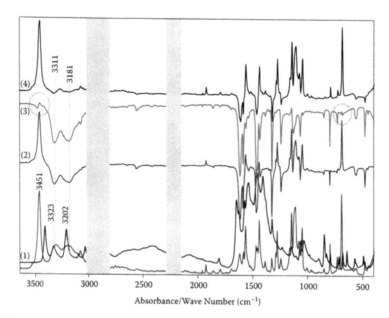

**FIGURE 4.21** Solid-state (1) nonpolarized IR, (2) difference IR-LD, and reduced IR-LD spectra in a nematic host of 2-chloro-3-aminopyridine after elimination of the bands at (3) 3451 cm⁻¹ and (4) 3311 cm⁻¹. IR spectrum of 2-chloro-3-aminopyridinium hydrogensquarate in nematic mesophase. Gray bars show the self-absorption band of the nematic mesophase.

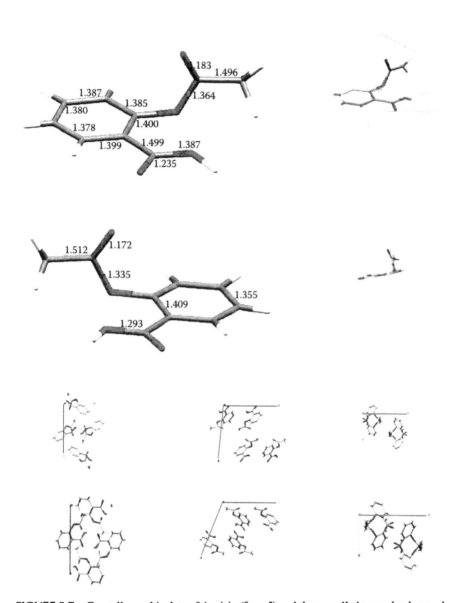

**FIGURE 5.7** Crystallographic data of Aspirin (form I) and the so-called second polymorph (form II) [368,379,380].

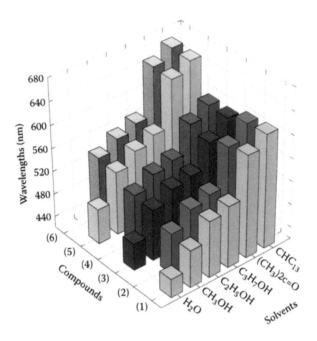

**FIGURE 6.8** Vis spectroscopic data for the aromatic and quinoide forms of 2-OH derivatives in different solvents.

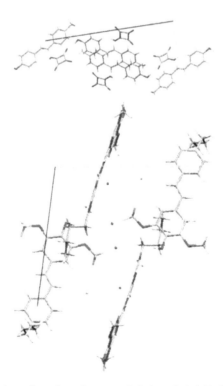

**FIGURE 6.15** View along the *a*, *b*, and *c* axes of (1) 1-methyl-4-[2-(4-hydroxyphenyl)ethenyl)]pyridinium] hydrogenphosphate and (2) 1-butyl-4-[2-(3,5-dimethoxy4-hydroxyphenyl) ethenyl)]piridinium] chloride tetrahydrate.

**FIGURE 6.15** (Continued)

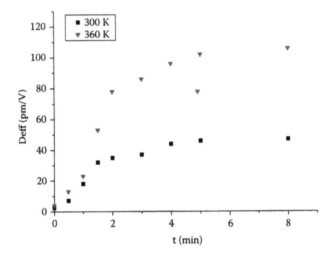

**FIGURE 6.20** Dependence of the second-order susceptibilities of 1-methyl-4-[2-(3-methoxy-4-hydroxyphenyl)ethenyl)]pyridinium] hydrogensquarate monohydrate at 1064 nm versus time with frequency repetition of about 1 kHz.

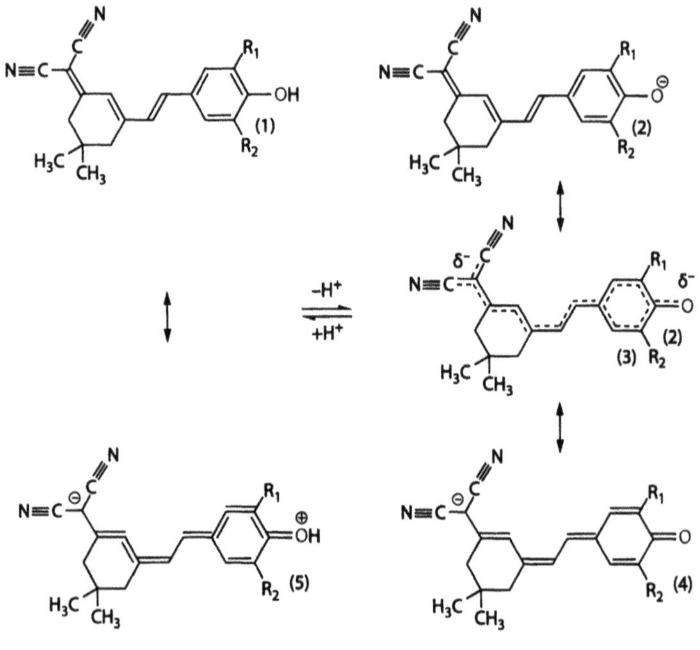

**FIGURE 6.29** Chemical diagram of different resonance forms compounds with $R_1 = R_2 = H$, $R_1 = H$, $R_2 = OCH_3$, and $R_1 = R_2 = OCH_3$.

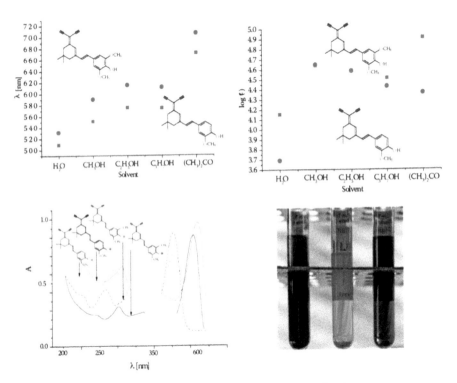

**FIGURE 6.30** Electronic spectra of different substituted dicyanoisophorones.

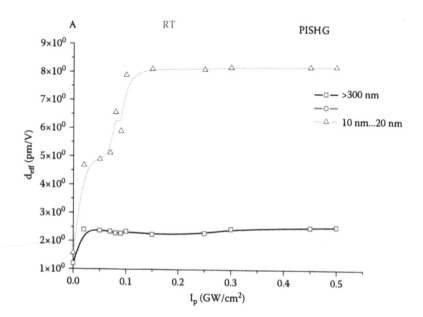

**FIGURE 6.32** Dependence of the second-order optical susceptibility for the sample 2-[3-[2-(4-hydroxyvinyl]-5,5-dimethylcyclohex-2-enylidene]malononitrile versus the effective photoinducing pump power during simultaneous treatment by two coherent laser beams with a ratio of the fundamental to the writing beam ratio equal to about 15.

# 5 Application in Pharmaceutical Analysis

The orientation technique of colloid suspensions in a nematic host has been successfully applied to the analysis of pharmaceutical products, especially for analysis of polymorphs or solid multicomponent mixtures (up to five components) [33]. In some of the described compounds, the method appears to be unique for experimental infrared (IR) band assignment. On this basis, quantitative approaches for the determination of pharmaceutical products in solid binary mixtures have been presented. An interesting practical field of application has been established in the analysis of the polymorphs of pharmaceutical products. We will now demonstrate the applicability of the method for the elucidation of some of the best-selling drugs (depicted in Figures 5.1 through 5.3).

The considerable commercial interest of the pharmaceutical industry in Paracetamol® (Sopharma, Bulgaria), polymorphs, and their quantitative analysis, as a result of their wide usage as antipyretic and analgetic agents, has led to the development of suitable methods of characterization. X-ray diffraction (XRD), for example, is a powerful technique for the identification of a crystalline solid phase. IR spectroscopy is also a recommended method.

Paracetamol (acetaminophen, 4′-hydroxyacetanilide) (Figure 5.1) is known as three known polymorphs: monoclinic (form I), orthorhombic (form II), and form III [343,344]. The first two are stable enough for experimental structural investigations and for that reason they are fully described in the literature [344–347]. The third form, obtained by melting of the monoclinic modification, is unstable. The phase transition of forms I and II under the effect of hydrostatic pressure and temperature has been studied [347–350]. The comparative studies of different polymorphs of Paracetamol and clarification of their physical and chemical properties are of great interest and importance because of their bioactivity and applicability in pharmacology.

For these reasons we have performed investigations by means of linear-dichroic infrared (IR-LD) spectroscopy analysis of oriented solids on both monoclinic and orthorhombic polymorphs of Paracetamol by the orientation technique as a liquid crystal (LC) suspension. The conventional IR spectral analysis of Paracetamol in solution has been previously demonstrated [351–355]. However, here for the presented CS-NLC orientation technique, IR-LD analysis leads to (a) detailed vibrational assignment of characteristic bands of both polymorphs; and (b) the supramolecular solid-state structural characterization at room temperature and atmospheric pressure. It also avoids the phase transition and guarantees the study of the different forms. A quantitative approach to the yield determination of monoclinic form I in mixtures with the orthorhombic modification was also illustrated.

Aspirin® (Bayer, Germany) (Figure 5.1), the world's best-selling drug, was synthesized in 1853 and was the subject of a series of papers dealing with the

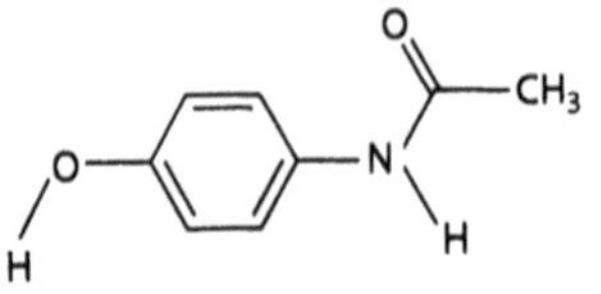

(1)

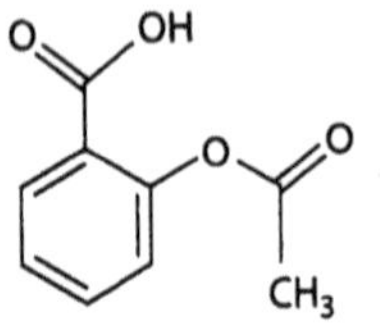

(2)

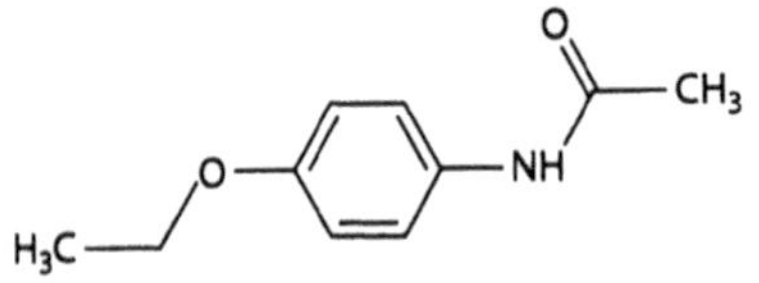

(3)

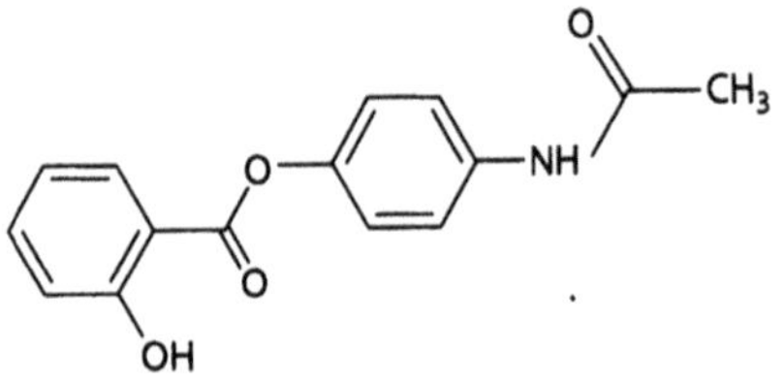

(4)

**FIGURE 5.1** Chemical diagrams for (1) Paracetamol, (2) Aspirin, (3) Phenacetin (1-acetamino-4-ethoxybenzol), and salophen ((4-acetamidophenyl) 2-hydroxybenzoate).

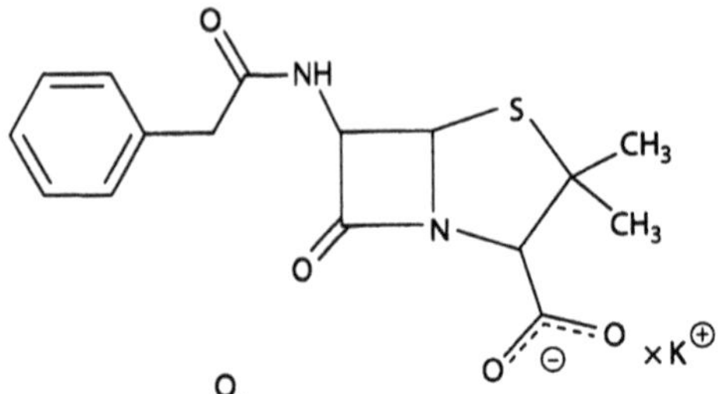

Benzylpenicillin

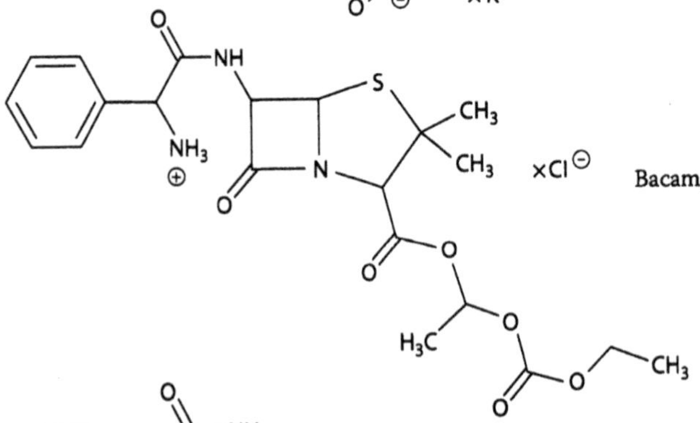

Bacampicillin

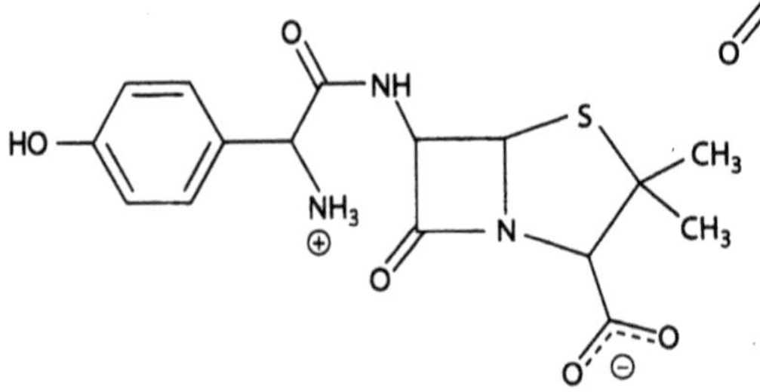

Amoxicillin

**FIGURE 5.2** Chemical diagrams for benzylpenicillin, bacampicillin, and amoxicillin.

prediction of different polymorph modifications [356–358]. The known form I was the subject of structural determination a long time ago; however, an elusive form II ("second polymorph") has been isolated by cocrystallization processes and was characterized by single-crystal X-ray structural analysis by Vishweshwar and coworkers in 2005 [359]. The report of this second polymorph created an intensive discussion in the literature concerning the exact refinement conditions. A redetermination of the structure under different conditions has been performed by Bond et al. [360,361]. This example may be the best one to demonstrate the applicability of the CS-NLC method for the analysis of the polymorphs.

Penicillins® (Merck & Co., Inc., United States) (Figure 5.2) are a group of well-known beta-lactam antibiotics used in the treatment of bacterial infections. Beginning with the discovery of penicillin by Scottish scientist Sir Alexander Fleming in 1928, though others had earlier noted the antibacterial effects of penicillium, 80 years later penicillin drugs are still used in pharmaceutical practice. Their wide usage as medication in therapy lead to an ever-growing interest in fast, simple, and reliable analytical methods for the determination of these pharmaceutical products in solid mixtures. As trade products, penicillin antibiotics are offered as tablets, powders for injection, or oral suspensions, and a number of quantitative methods for their determination in solution have been reported [362–366]. Solid-state physical properties of amoxicillin have been investigated, depending on the pH values by means of powder XRD measurements [367]. The solid-state fluorescence of this compound has been described [368]. Analysis of amoxicillin in human urine by photoactivated generation of fluorescence excitation–emission matrices has also been presented [367]. A Raman study combined with density functional theoretical calculations of potassium benzylpenicillin on silver colloids at different pH values has been reported [369]. However, in the solid state, most of these antibiotics are characterized by a broad fluorescence band leading to the interpretation of the Raman spectra and quantitative determination in solids being possible. For these reasons, polarized IR-LD spectroscopy was employed to develop a method for quantification of three examples of the penicillin drugs: 3,3-dimethyl-6-oxo-7-(2-phenylacetyl)amino-2-thia-5-azabicyclo[3.2.0]heptane-4-carboxylic acid (benzylpenicillin, penicillin G); 1-ethoxycarbonyloxyethyl (2S,5R)-6-[[(2R)-2-amino-2-phenyl-acetyl]amino]-3,3-dimethyl-7-oxo-4-thia-1-azabicyclo[3.2.0]-heptane-2-carboxylate hydrochloride (bacampicillin); and (2S,5R,6R)-6-[(R)-(-)-2-amino-2-(p-hydroxyphenyl)acetamido]-3,3-dimethyl-7-oxo-4-thia-1-azabicyclo[3.2.0]heptane-2-carboxylic acid trihydrate (amoxicillin) in solid binary mixtures. The analysis of these systems is of significant interest in pharmaceutical practice. The determination of penicillin G and amoxicillin in a mixture with cephalexin by means of liquid chromatography and second derivative spectroscopy has been presented [370,371]. However, to the best of our knowledge a quantitative approach to the determination of the penicillins in the solid state by means of IR spectroscopy had not been described in the literature prior to our study.

Cephalosporins (Figure 5.3) have antibactericidal activity and exhibit the same mode of interaction as other beta-lactam antibiotics such as penicillin derivatives [372,373]. Cephalosporins disrupt the synthesis of the peptidoglycan layer of

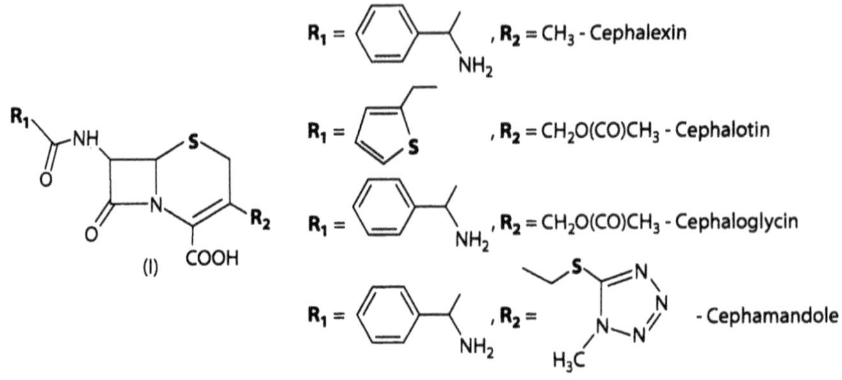

$R_1 =$ ⬡—< , $R_2 = CH_3$ - Cephalexin
$\overset{|}{NH_2}$

$R_1 =$ (thiophene)—< , $R_2 = CH_2O(CO)CH_3$ - Cephalotin

$R_1 =$ ⬡—< , $R_2 = CH_2O(CO)CH_3$ - Cephaloglycin
$\overset{|}{NH_2}$

$R_1 =$ ⬡—< , $R_2 =$ (tetrazole-thio) - Cephamandole
$\overset{|}{NH_2}$

**FIGURE 5.3** Chemical diagram of cephalosporin nucleus (**I**) and compounds studied: cephalexin hydrochloride monohydrate, cephalotin sodium salt, cephaloglycin hydrochloride, and cephamandole sodium salt.

bacterial cell walls. The peptidoglycan layer is important for cell wall structural integrity, especially in Gram-positive organisms. The final transpeptidation step in the synthesis of the peptidoglycan is facilitated by transpeptidases known as penicillin binding proteins. Their wide usage as medication in therapy led to an ever-growing interest in fast, simple, and reliable analytical methods for the determination of these pharmaceutical products in binary mixtures. The quantitative methods for determination of cephalosporins in liquid binary mixture were developed in the 1990s. Two-component mixtures of cephapirin and cefuroxime salts were assayed by Morelli [374] by first- and second-derivative spectrophotometric methods using a "zero-crossing" technique of measurement. The resolution of binary mixtures of cephalothin and clavulanic acid by using first-derivative spectrophotometry has been discussed. Usually cephalosporin antibiotics are offered as trade products as a powder mixture for injection. In these cases powder X-ray diffraction is a powerful and routine method for the identification of mixtures.

The difference IR-LD spectrum of the monoclinic Paracetamol form I (Figure 5.4) is characterized with a positive peak at 3324 cm$^{-1}$ ($v_{NH}$) and the absence of a respective one at 3160 cm$^{-1}$ ($v_{OH}$). The peak positions indicate the presence of strong intermolecular interactions, as established by XRD, that shows two types of intermolecular hydrogen bonds—OH⋯O=C (2.663Å) and NH⋯OH (2.934Å)—thus forming pleated layers with a torsion angle between the phenyl fragment and amide group of 21.1°. An intensive positive $v_{C=O}$ band (Amide I) at 1656 cm$^{-1}$ and a series of negative maxima at 1610 cm$^{-1}$ (**8a**, $A_1$), 1565 cm$^{-1}$ ($\delta_{NH}$, Amide II), and 1508 cm$^{-1}$ (**19a**, $A_1$) are similar to those obtained for other p-substituted N-arylacetanilides [375–377] (Figure 5.4). The characteristic IR regions 1260 to 1200 cm$^{-1}$ and 840 to 750 cm$^{-1}$ for form I contain (a) three intensive negative bands at 1259 cm$^{-1}$, 1243 cm$^{-1}$, and 1226 cm$^{-1}$ of **9a** ($A_1$ in-plane [i.p.] *Ph*), $v_{C-O(H)}$ stretching, and $v_{PhN}$; and (b) negative peaks at 836 cm$^{-1}$, 808 cm$^{-1}$, and 796 cm$^{-1}$ of 11-$\gamma_{CH}$, 4-$\gamma_{ArH}$ (out-of-plane [o.p.] modes) of the p-substituted aromatic system and **12(X)** ($A_1$ i.p.).

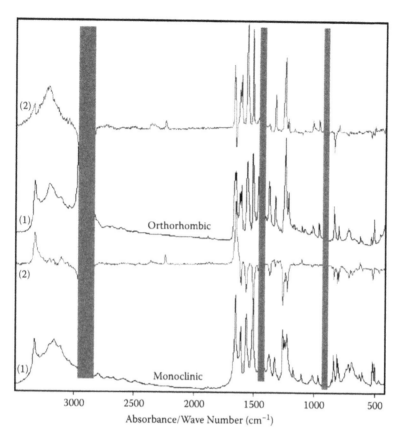

**FIGURE 5.4**   (1) Nonpolarized IR and (2) difference IR-LD spectra of monoclinic and orthorhombic polymorphs of Paracetamol. Self-absorption regions of the NLC are given in gray rectangles.

Application of the reducing-difference procedure (RDP) leads to the following results: (a) the elimination of the peak of $v_{NH}$ (3324 cm$^{-1}$) causes the disappearance of the $v_{C=O}$ peak (1656 cm$^{-1}$) (Figure 5.5), thus confirming the X-ray results for planar and endo trans-configuration of the amide fragment (Figure 5.6); and (b) the elimination of the **8a** peak at 1610 cm$^{-1}$ leads to vanishing of the 1506 cm$^{-1}$, 1259 cm$^{-1}$, 1243 cm$^{-1}$, 1227 cm$^{-1}$, 836 cm$^{-1}$, 808 cm$^{-1}$, 798 cm$^{-1}$, 690 cm$^{-1}$, and 520 cm$^{-1}$ peaks, and reduction of the 1376 cm$^{-1}$, 1327 cm$^{-1}$, 1376 cm$^{-1}$ peaks. These results correlate with the XRD data as well as with group analysis, indicating a colinear orientation of $A_1$ i.p. and $B_1$ o.p. modes of phenyl rings. The simultaneous elimination of both mode types is impossible within the framework of one phenyl fragment, as a result of their mutually perpendicular orientation. On the other hand, this result also confirms the previously listed frequency assignment; (a) those at the 1259 cm$^{-1}$, 1243 cm$^{-1}$, and 1227 cm$^{-1}$ peaks to the **9a**, $v_{C-O(H)}$, and $v_{PhN}$ vibrations, respectively; (b) the character of the 1371 cm$^{-1}$, 1328 cm$^{-1}$, and 520 cm$^{-1}$ modes shows their attribution to $\delta^s_{CH3}$, Amide III ($v_{C-N}$) and $\gamma_{C=O}$, as far as the corresponding transition moments

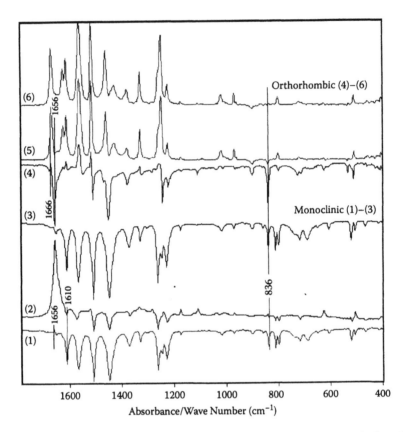

**FIGURE 5.5** (1) Difference and reduced IR-LD spectra of Paracetamol form I after elimination of (2) 1610 cm$^{-1}$, and (3) 3324 cm$^{-1}$ peaks. (4) Difference and reduced IR-LD spectra of Paracetamol form II after elimination of (5) 1666 cm$^{-1}$ and (6) 836 cm$^{-1}$ peaks.

are colinearly oriented toward the $B_1$ o.p. and $A_1$ i.p. directions; and (c) the wide absorption maximum, which is also reduced in the spectrum of Figure 5.4 at 690 cm$^{-1}$, is assigned to the $\gamma_{NH}$ mode like the other p-substituted N-arylacetanilides. However, the detailed interpretation of the series of bands between 840 cm$^{-1}$ and 760 cm$^{-1}$ is complicated as a result, illustrating the perpendicular disposition of the two Paracetamol molecules in the unit cell of monoclinic form with a pleated sheet stacked along the crystallographic $b$ axis. However, the exact assignment is possible and is demonstrated in the following IR-LD analysis of orthorhombic II form (Figure 5.6).

The orthorhombic polymorph II is characterized by flat layers and possesses parallel hydrogen bonded sheets along the $c$ axis. Intermolecular hydrogen OH···O=C (2.742 Å) and NH···OH (2.967 Å) bonds are found as well as a deviation from a coplanar disposition of the amide group and phenyl ring of 17.7°. The C=O bond is shorter than the analogous one in the monoclinic polymorph and causes the higher frequency shifting of the $v_{C=O}$, $v_{NH}$, and $v_{OH}$ bands. The two-dimensional (2D) hydrogen-bonded networks in both polymorphs can also be considered as being built from

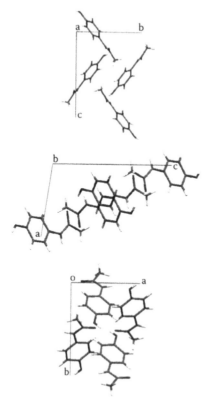

**FIGURE 5.6** Crystallographic data of monoclinic Paracetamol viewed along the *a*, *b*, and *c* axes.

chains of Paracetamol molecules. On defining the orientation of individual chains in the sequence of chains within their layers can be described as A-A-A in polymorph I and A′-B-A′ in II, where A and B differ in the orientation, while A′ and A have the same orientation of the molecules but the intermolecular bonds and angles are different.

In contrast to the IR spectrum of the monoclinic form, the solid-state IR spectrum of the orthorhombic form contains the following main differences: (a) the $v_{OH}$ band is shifted to a higher frequency (3205 cm$^{-1}$); (b) the 1750 to 1450 cm$^{-1}$ region of form II has two intensive maxima at 1666 cm$^{-1}$ and 1656 cm$^{-1}$; (c) two peaks at 1610 cm$^{-1}$ and 1623 cm$^{-1}$; (d) low-frequency shifting of the Amide II peak at 1558 cm$^{-1}$; (e) an intensive peak at 1243 cm$^{-1}$ in the 1270 to 1170 cm$^{-1}$ region and a low frequency shifted one at 1220 cm$^{-1}$ that are accompanied with a shoulder at 1259 cm$^{-1}$; (f) the absence of the 808 cm$^{-1}$ mode in the 870 to 760 cm$^{-1}$ region; and (g) the high frequency shifting of the 520 cm$^{-1}$ peak to 530 cm$^{-1}$ in form II. The first result correlates with the XRD data, and is due to longer, intermolecular OH⋯O(=C) bonds compared to the monoclinic form, leading to the observed high frequency shifting of the $v_{OH}$ band. The other changes can in many case be explained as resulting from crystal field splitting or different crystal packing in the polymorphs.

The difference spectrum of Paracetamol form II (Figure 5.5) has a reduced band at 3326 cm$^{-1}$ ($v_{NH}$) and a positive maximum at 3205 cm$^{-1}$ ($v_{OH}$). The 1500 to 1700 cm$^{-1}$ region is characterized with a positive series of maxima at 1666 cm$^{-1}$, 1610 cm$^{-1}$, 1623 cm$^{-1}$, 1558 cm$^{-1}$, and 1513 cm$^{-1}$, and a negatively oriented peak at 1656 cm$^{-1}$. The different orientation of both the 1666 cm$^{-1}$ and 1656 cm$^{-1}$ peaks (Figure 5.5) rejects the eventual crystal field splitting of the $v_{C=O}$ peak, which presumably would lead to an equal orientation of both bands at 1666 cm$^{-1}$ and 1656 cm$^{-1}$. The positive peaks at 1245 cm$^{-1}$, 1218 cm$^{-1}$, and 798 cm$^{-1}$, and the negative ones at 836 cm$^{-1}$ and 530 cm$^{-1}$ are also determined. In this case the $A_1$ modes of one layer are colinearly oriented toward **n** and positive in the difference spectrum in contrast to negative $B_1$ o.p. ones. The NH bond is placed at an angle near to the magic value (54.7°) and the corresponding peak in the difference spectrum is practically eliminated.

The elimination of the 1666 cm$^{-1}$ peak (Figure 5.5) leads to the disappearance of the 3205 cm$^{-1}$ ($v_{OH}$), 1623 cm$^{-1}$, 1610 cm$^{-1}$, 1513 cm$^{-1}$, 1257 cm$^{-1}$, 721 cm$^{-1}$, and 454 cm$^{-1}$ peaks. The simultaneous elimination of both the 1623 cm$^{-1}$ and 1610 cm$^{-1}$ peaks confirms their doublet character as a result of the crystal field splitting of the **8a** mode and indicates the assignment of the 721 cm$^{-1}$ peak to the $\delta_{C=O}$ i.p. mode. The same procedure applied to the 836 cm$^{-1}$ peak leads to the disappearance of the broad peaks at 700 cm$^{-1}$ and 530 cm$^{-1}$, thus showing that they belong to 11-$\gamma_{CH}$ o.p., $\gamma_{NH}$ and $\gamma_{C=O}$, respectively. Therefore, the peak at 798 cm$^{-1}$ corresponds to the 12 $A_1$ i.p. mode, which is identical with the analogous one in the monoclinic form. The last elimination was provoked in the 1700 to 1000 cm$^{-1}$ region and the disappearance of only the 1565 cm$^{-1}$ maximum indicates its belonging to combination bands (836 cm$^{-1}$ + 817 cm$^{-1}$). Moreover, the 1565 cm$^{-1}$ and 836 cm$^{-1}$ peaks exhibit the same orientation in the different IR-LD spectrum. These results also illustrate the assignment of the 808 cm$^{-1}$ peak in form I to the **4**-$\gamma_{ArH}$ o.p. mode, whose intensity is very sensitive to the crystal packing and for this reason has low but adequate intensity in form II, but to determine its simultaneous elimination with the 836 cm$^{-1}$.

The ratios of the characteristic peaks of one form (1666 cm$^{-1}$, form II) with the second $A_1$ mode at 1610 cm$^{-1}$ present in both polymorphs were considered for quantification of the two modifications in mixtures using the corresponding model [378]. The significant intensity of both of the discussed bands, compared to the 806 cm$^{-1}$ and 836 cm$^{-1}$ bands, suggested that an improvement of corresponding quantitative approach would be possible. Repeated IR spectroscopic analysis of three replicated samples for each mole fraction was applied and the results of the mean peak ratio are presented. Linear regression analysis between content and the peak ratio data gave a straight-line plot: $y = 0.03$ ($\pm 0.04$)•$x + 0.87$ ($\pm 0.07$), where $x$ is the mole fraction of the monoclinic form ($x = 1/X_1$). Confidence intervals from 0.61 to 0.48 and 0.79 to 0.47 for slope and intercept are obtained. Therefore, the linear relationship is $y = 0.034$•$x + 0.875$. The corresponding $r$ factor is 0.9974 for the calibration points on the regression line. The square quadratic value ($r^2$) of $r$ is equal to 0.9948, which gives a reliability index of 99.48% for the quantitative method. This good result is, similar to the reliability indices of known quantitative methods [378] and confirms the precision and the accuracy of the Fourier transform infrared (FT-IR) spectroscopic determination of monoclinic Paracetamol in mixtures with the orthorhombic modification.

One of the best examples of the successful application of polarized IR spectroscopy for analysis of the different polymorphs of pharmaceutical products is that of Aspirin and its pseudo "polymorph" modification (Figure 5.7). According to the crystallographic data, form I crystallizes in the $P2_1/c$ space group and consists of centrosymmetric carboxylic acid dimer moieties (O···O: 2.635 Å, 177.7°) that are, in turn, joined via centrosymmetric methyl CH···O (C···O: 3.553 Å, 164.0°) contacts of acetyl groups, thus forming 1D chains [360,361] (R factor 7.97%). The crystallographic data of form II (R factor 16.22%) indicated the formation of intermolecular OH···O=C and HNH···O=C bonds with participation of carboxylic COOH and primary amide O=C-NH$_2$ groups at bond lengths of 2.913 and 2.565 Å [360,361].

We performed polarized IR-LD investigations on forms I and II and will now illustrate how IR-LD spectroscopy can relatively easily provide information on the polymorphism of the pharmaceutical products.

The difference IR-LD spectrum of Aspirin form I illustrated a significant (Figure 5.8) level of macroorientation of the solid guest molecule in the nematic liquid crystal (NLC) mesophase. The broad absorption peak between 3200 and 2300 cm$^{-1}$ belongs to the $v_{OH}$ stretching vibration of COOH group in the Aspirin molecule and indicated the presence of strong intermolecular interactions as confirmed by single-crystal X-ray diffraction. The higher frequency band in the 1800 to 1400 cm$^{-1}$ region at 1752 cm$^{-1}$ belongs to the $v_{C=O}$ stretching mode of the acetoxy group and the maximum at 1693 cm$^{-1}$ to the $v_{C=O}$ mode of a carboxylic one as a result of their different intermolecular hydrogen bond lengths. Both maxima are observed as doublets, which could be explained with the crystal field splitting effect. The adequate experimental band assignment of the IR bands in the spectral region lower than 1400 cm$^{-1}$ is possible through the following IR-LD spectral analysis. Elimination of the 1752 cm$^{-1}$ and 1693 cm$^{-1}$ peaks at different dichroic ratios is a result of the near to perpendicular orientation of both $v_{C=O}$ transition moments in the frame of the Aspirin molecule. They are inclined at a dihedral angle of 83.6° (Figure 5.7). Elimination of the 1606 cm$^{-1}$ peak provoked the disappearance of peaks at 1307 cm$^{-1}$ and 917 cm$^{-1}$ (Figure 5.8) possessing the same symmetry class (A$_1$) and assigned as **3** and **1** modes of the o-disubstituted benzene ring (Figure 5.9). The simultaneous disappearance of the 838 cm$^{-1}$ and 755 cm$^{-1}$ peaks assigned their out-of-plane origin (B$_1$ symmetry class) and then belonging to **11**-$\gamma_{CH}$ and **4**-$\gamma_{Ar}$ bending vibrations. Reduction of peaks at 668 cm$^{-1}$ and 599 cm$^{-1}$ at the same dichroic ratio, assigned their origin as **12(X)** and **6a(X)**, which are both in-plane bending modes.

The cocrystallization of Aspirin with acetamide and formation of form II resulted in a complication of the IR spectral curve (Figure 5.9) as a result of the strong self-absorption of the cocrystallizing component. The 3500 to 2500 cm$^{-1}$ region is characterized with four absorption peaks at 3365 cm$^{-1}$ ($v^{as}_{NH2}$), 3260 cm$^{-1}$ ($v_{OH}$), and the FR doublet about $v^s_{NH2}$ at 3305 cm$^{-1}$ and 3175 cm$^{-1}$ obtained by preliminary deconvolution and curve-fitting procedure (CFP). The last assignment is based as well on the known crystallographic data, which established the formation of intermolecular OH···O=C and HNH···O=C bonds in form II with participation of carboxylic COOH and primary amide O=C-NH$_2$ groups with bond lengths at 2.913 and 2.565 Å. The 1700 to 1400 cm$^{-1}$ region is also complicated and being strongly overlapped also requiring a preliminary deconvolution and CFP treatment to a series of peaks at

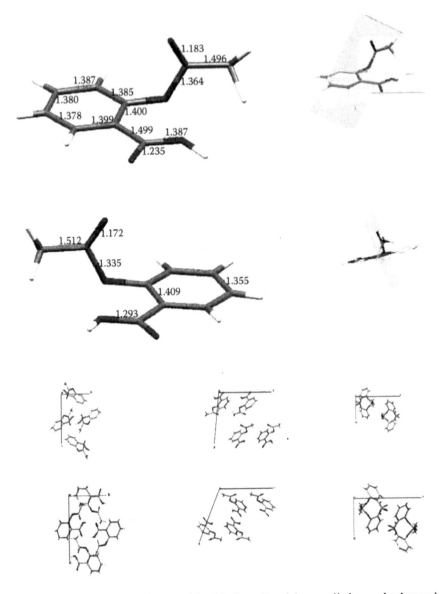

**FIGURE 5.7** Crystallographic data of Aspirin (form I) and the so-called second polymorph (form II) [368,379,380]. **(See color insert.)**

1751 cm$^{-1}$, 1695 cm$^{-1}$, 1680 cm$^{-1}$, 1654 cm$^{-1}$, and 1606 cm$^{-1}$ assigned to $v_{C=O}$ (acetoxy-group), $v_{C=O}$ (carboxylic group), $v_{C=O}$ (Amide I), and $\delta_{NH2}$ (Amide I) modes, respectively. Like form I, the doublet character of first two maxima could be explained with the crystal field splitting effect.

A significant level of macroorientation of the guest molecule is also established in form II (Figure 5.8). The elimination of Amide I mode at 1680 cm$^{-1}$ resulted in the

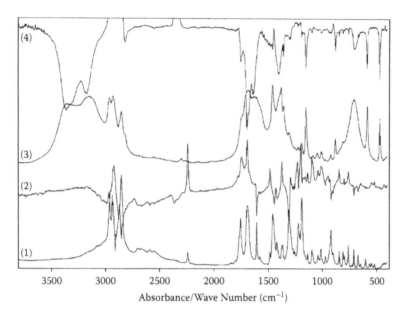

**FIGURE 5.8**   (1 and 3) Nonpolarized IR and (2 and 4) difference IR-LD spectra of Aspirin polymorphs (form I [434,435] and pseudo form II).

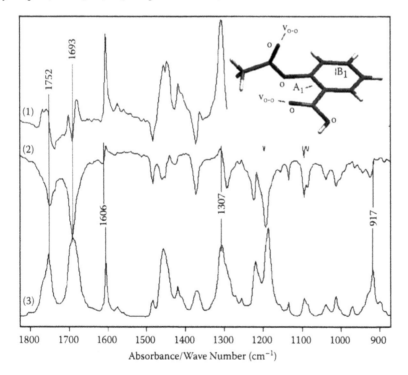

**FIGURE 5.9**   (1) Nonpolarized IR and reduced IR-LD spectra of form II of Aspirin after elimination of the bands at (2) 1606 cm$^{-1}$ and (3) 1693 cm$^{-1}$.

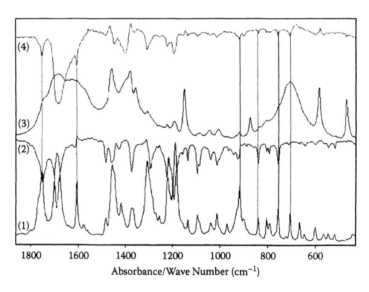

**FIGURE 5.10**  Nonpolarized IR spectra of (1) form I and (3) form II of Aspirin as well as reduced IR-LD spectra of (2) form I after elimination of the band at about 1600 cm⁻¹ and (4) of form II after the elimination of the characteristic IR bands of the cocrystallized component.

disappearance of the 1751 cm⁻¹ peak as a result of colinear orientation of corresponding transition moments in the framework of the unit cell (Figure 5.10). Typically, the asymmetric intermolecular interactions of the amide $NH_2$ group influence the character of the $\gamma_{NH2}$ mode peak within the 740 and 801 cm⁻¹ region with a maximum at 725 cm⁻¹, and strongly overlapping o.p. maxima at 836 cm⁻¹ and 755 cm⁻¹. Their simultaneous elimination occurred during the last procedure, which is also in accordance with single-crystal X-ray data indicating a colinearity of the $\nu_{C=O}$ of acetoxy group, Amide I, and out-of-plane modes of the 1,2-disubstituted benzene fragment in the framework of the unit cell of polymorph form II. The other characteristic maxima are insignificant because the molecule geometry of Aspirin in form II is slightly different in terms of torsion angles defined by carboxylic and acetyl groups.

However, the elimination of the characteristics for the IR bands of cocrystallizing compound generates the IR maxima of Aspirin in form II (Figure 5.10). The observed IR bands coincide with those of the form I. Using the method for the identification of compounds by means of their integral absorbence ratios, we analyzed both the forms. The data of the bands at 842, 804, and 756 cm⁻¹ (Figure 5.10 and Figure 5.11) are summarized in Table 5.1. Coincidences of 98% and 94% are obtained, which is in accordance with the conclusions about the nature of the so-called second polymorphs [360, 361]. Regarding these data, there still exists only one polymorph of Aspirin, looking at the commonly accepted definition of polymorphism.

The data from the IR-LD analysis of cephalosporins, that is, *cephalexin*, 8-(2-amino-2-phenyl-acetyl)amino-4-methyl-7-oxo-2-thia-6-azabicyclo [4.2.0] oct-4-ene-5-carboxylic acid; *cephalotin*, ((6R,7R)-3-(acetoxymethyl)-8-oxo-7-(2-(thiophen-2-yl)acetamido)-5-thia-1-aza-bicyclo[4.2.0]oct-2-ene-2-carboxylic

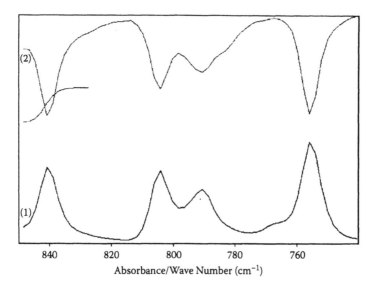

**FIGURE 5.11**  Integration of selected IR bands in (1) form I and (2) form II.

**TABLE 5.1**
**Integral Absorbence Ratios ($A_{v_1}/A_{v_2}$) of Selected IR Bands in Forms I and II of Aspirin**

| Form | $A_{842}/A_{756}$ | $A_{842}/A_{804}$ |
|------|-------------------|-------------------|
| I    | $0.8960_1$        | $0.6135_7$        |
| II   | $0.8757_2$        | $0.6523_4$        |

acid, (2)); *cephaloglycin*, (4-(acetoxymethyl)-8-(2-amino-2-phenyl-acetyl)amino-7-oxo-2-thia-6-azabicyclo[4.2.0]oct-4-ene-5-carboxylic acid; and *cephamandole*, ((6R,7R)-7-{[2R)-2-hydroxy-2-phenylacetyl]amino}-3-[(1-methyltetrazol-5-yl) sulfanylmethyl]-8-oxo-5-thia-1-azabicyclo[4.2.0]oct-2-ene-2-carboxylic acid are summarized in Table 5.2. The quantitative IR and Raman spectroscopic approach to the determination of four cephalosporin antibiotics in six binary solid mixtures is presented. Previously, only quantitative tools for determination of cephalosporins in solution were reported. Absorbance ratios of the bands at 800 cm$^{-1}$, characteristic for each of the determined compounds to the bands at about 721 cm$^{-1}$ present in each of the cephalosporins, that is, cephalexin, cephalotin, cephaloglycin, and cephamandole, were used. The $r^2$ values give confidence levels within the 99.32%–99.88% and 99.90%–95.54% ranges for the systems studied using the absorption ratios of the bands at 800 cm$^{-1}$ and 720 cm$^{-1}$ in IR and Raman spectra, respectively. Solid-state IR-LD spectral analysis of solid mixtures was carried out with the goal of obtaining experimental IR spectroscopic assignment of the characteristic IR bands of all of the

## TABLE 5.2
### Selected IR and Raman Bands of Cephalexin, Cephalotin, Cephaloglycin, and Cephamandole in the Solid State

| | $\nu$ [cm$^{-1}$] | | | | | | | |
|---|---|---|---|---|---|---|---|---|
| | Cephalexin | | Cephalotin | | Cephaloglycin | | Cephamandole | |
| Assignment | IR | R | IR | R | IR | R | IR | R |
| $\nu_{OH\ (H2O)}$ | 3619 | 3549 | — | — | — | — | — | — |
| $\nu_{OH}$ | 3550 | — | — | — | — | — | 3563 | — |
| $\nu_{NH}$ | 3369 | 3369 | 3282 | 3280 | 3303 | 3303 | 3336 | 3336 |
| $\nu_{C=Ocomb.}$ | 1776 | 1777 | 1751 | 1750 | 1776 | 1777 | 1758 | 1760 |
| | 1756s | 1756w | 1729 | 1730w | 1751 | 1751w | 1731 | 1761w |
| $\nu_{C=O(ester)}$ | — | — | 1743 | — | 1745 | 1745w | — | — |
| $\nu_{C=O\ (Amide\ I)}$ | 1670s | 1670w | 1660 | 1660w | 1689 | 1689w | 1679s | 1679w |
| $\nu^{as}_{COO-}$ | — | — | 1623 | 1623w | — | — | 1652 | 1651 |
| **8a, 8b** | 1596 | 1596 | — | — | 1616 | 1615 | 1610 | 1610 |
| | 1575 | 1575 | | | 1577 | 1577 | 1592 | 1592 |
| $\delta_{NH\ (Amide\ II)}$ | 1538 | 1537 | 1530 | 1530 | 1558 | 1558w | 1525 | 1523 |
| **19a** | 1496 | 1495 | — | — | 1500 | 1500 | 1510 | 1510 |
| $\delta_{C=O\ (Amide\ IV)}$ | 800 | 800 | 800 | 799 | 800 | 800 | 800 | 801 |
| $11\text{-}\gamma_{CH}$ | 759 | 759 | — | — | 755 | 755 | 765 | 766 |
| $4\text{-}\gamma_{Ar}$ | 698 | 698 | — | — | 696 | 697 | 701 | 701 |
| $\nu^s_{CSC}$ | 721 | 721 | 742 | 741 | 738 | 738 | 723 | 723 |
| $\gamma_{C=O}$ | 582 | 581 | 578 | 579 | 576 | 576 | 574 | — |

investigated compounds. The CS-NLC orientation technique was used, combined with RDP for polarized spectra interpretation. High-performance liquid chromatography with tandem mass spectrometry, using electrospray ionization (HPLC-ESI-MS/MS) analysis was performed independently for comparisons and validation of vibrational spectroscopy data. Applications to 10 tablets of the commercial products Cefamandole® and Cefalotin® (Actavis) were reported.

On the basis of IR and Raman spectroscopy (Table 5.2, Figure 5.12), which are relatively cheap, fast, and easy for technical operation and interpretation of data, and do not require sample dissolution methods, we presented the quantitative determination of the six binary mixtures with the studied cephalosporins in solid state, which was reported for the first time in the literature. The IR-LD analysis of oriented colloids as a liquid crystal suspension was applied for experimental IR band assignment and selection of appropriate bands for quantitative determination. This method gives additional supramolecular solid-state structural information at room temperature and atmospheric pressure. It also avoids the phase transition and guarantees the study of different forms without polymorph transitions. This approach has been applied recently for caffeine as a matrix compound and for studying the polymorphs of Paracetamol, Aspirin, Phenacetin®, and Salophen®. The spectroscopic data were

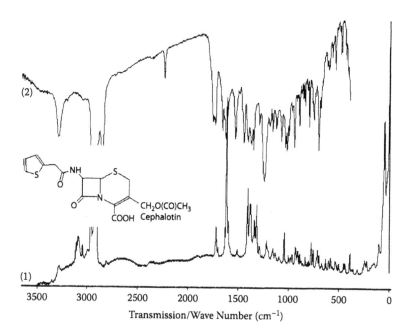

**FIGURE 5.12**   (1) Raman and (2) IR spectra of Cephalotin in the solid state.


compared to HPLC-ESI-MS results for the quantitative amounts of the systems studied. The obtained correlations were used for determination of three trade products of powder samples of Cefamandole and Cefalotin.

For a two-component mixture containing the components $i$ and $j$ at concentrations $c_i$ and $c_j$, the total absorbance $A^t$ of the mixture at a given frequency is the sum of the absorbances of the compounds, $i$ and $j$, at the specified frequency: $A^t = A^i + A^j = a^i \times b \times c_i + a^j \times b \times c_j$, where $b$ is the sample thickness. Molar absorbtivities $a^i$ and $a^j$ are determined from absorption measurements of mixtures containing known amounts of compounds $i$ and $j$ at two different frequencies, $v_1$ and $v_2$. If the ratio of total absorbance of given band at frequency $v_1$ ($A^t_{v1}$) to the absorbance of a second band at frequency ($A^t_{v2}$), typical for only one component of mixture is determined, then the following equation can be formulated: $A^t_{v1}/A^i_{v2} = [(a^i_{v1} - a^j_{v1})/a^i_{v2}] + (a^j_{v1}/a^i_{v2}) \times (1/X_i)$ where $X_i = c_i/(c_i + c_j)$ and $A^j_{v2} = A^t_{v2}$.

Suitable frequencies for this purpose are:

System *(Cephalexin)/(Cephalotin)*: $v_1 = 800$ cm⁻¹ and $v_2 = 721$ cm⁻¹
System *(Cephalexin)/(Cephaloglycin)*: $v_1 = 800$ cm⁻¹ and $v_2 = 721$ cm⁻¹
System *(Cephalexin)/(Cephamandole)*: $v_1 = 800$ cm⁻¹ and $v_2 = 721$ cm⁻¹
System *(Cephalotin)/(Cephaloglycin)*: $v_1 = 800$ cm⁻¹ and $v_2 = 738$ cm⁻¹
System *(Cephalotin)/(Cephamandole)*: $v_1 = 800$ cm⁻¹ and $v_2 = 723$ cm⁻¹
System *(Cephaloglycin)/(Cephamandole)*: $v_1 = 800$ cm⁻¹ and $v_2 = 723$ cm⁻¹

In addition the *(Cephalexin)/(Cephalotin)* system is determined using the $v_1 = 800$ cm$^{-1}$ and $v_2 = 1756$ cm$^{-1}$ in the corresponding IR spectra.

Similar equations are used for Raman measurements [378]. For adequate comparison of the reliability of the IR and Raman data, the same bands are utilized for quantitative analysis. In corresponding Raman spectra, the character of the used bands is the same, but their relative intensity is different to the relative absorbance of the IR bands.

The ratios of the aforementioned IR characteristic peaks, typical for both compounds in corresponding solid mixtures, have been evaluated for quantification using the aforementioned mathematical model. Repeated IR and Raman analyses of samples (three replicates) for each mole fraction and each system were applied. Linear regression analysis between contents and peak ratio data gave straight-line calibrations listed in Table 5.3 and Table 5.4. The corresponding correlation coefficients $r$ and $r^2$ are given in Table 5.3 as well. In all cases, $p < 0.0001$ values indicate significant correlation. The squared values $r^2$ give confidence intervals within 99.88% and 99.32% and 99.90% and 95.54% for the systems studied using the absorption ratios of the bands at about 800 cm$^{-1}$ and 720 cm$^{-1}$ in IR and Raman spectra, respectively. For the case of the system *(Cephalexin)/(Cephalotin)* a straight-line plot $y = -19.020$

---

**TABLE 5.3**

**Straight-Line Calibrations, Where $y = A^t_{v1}/A^i_{v2}$ and $x = 1/X_i$, and $r$ Values for the Six Systems as Obtained by the IR-Spectroscopic Method**

| System | Straight-Line Calibrations | $r$ |
|---|---|---|
| *(Cephalexin)/(Cephalotin)* | $y = -15.680\ (\pm 0.96) + 14.14\ (\pm 0.23) \times x$ | 0.9984 |
| *(Cephalexin)/(Cephaloglycin)* | $y = 0.067\ (\pm 0.01) + 0.020\ (\pm 0.01) \times x$ | 0.9994 |
| *(Cephalexin)/(Cephamandole)* | $y = 0.521\ (\pm 0.03) + 0.354\ (\pm 0.01) \times x$ | 0.9991 |
| *(Cephalotin)/(Cephaloglycin)* | $y = -0.882\ (\pm 0.43) + 3.337\ (\pm 0.11) \times x$ | 0.9966 |
| *(Cephalotin)/(Cephamandole)* | $y = -13.559\ (\pm 0.85) + 12.500\ (\pm 0.21) \times x$ | 0.9991 |
| *(Cephaloglycin)/(Cephamandole)* | $y = -0.807\ (\pm 0.14) + 1.370\ (\pm 0.03) \times x$ | 0.9979 |

---

**TABLE 5.4**

**Straight-Line Calibrations, Where $y = I^t_{v1}/I^i_{v2}$ and $x = 1/X_i$, and $r$ Values for the Six Systems as Obtained by the Raman-Spectroscopic Method**

| System | Straight-Line Calibrations | $r$ |
|---|---|---|
| *(Cephalexin)/(Cephalotin)* | $y = -14.130\ (\pm 0.69) + 13.39\ (\pm 0.17) \times x$ | 0.9995 |
| *(Cephalexin)/(Cephaloglycin)* | $y = 0.081\ (\pm 0.01) + 0.012\ (\pm 0.01) \times x$ | 0.9933 |
| *(Cephalexin)/(Cephamandole)* | $y = 0.564\ (\pm 0.05) + 0.326\ (\pm 0.01) \times x$ | 0.9956 |
| *(Cephalotin)/(Cephaloglycin)* | $y = -0.635\ (\pm 0.38) + 3.210\ (\pm 0.09) \times x$ | 0.9972 |
| *(Cephalotin)/(Cephamandole)* | $y = -11.157\ (\pm 1.37) + 11.340\ (\pm 0.33) \times x$ | 0.9971 |
| *(Cephaloglycin)/(Cephamandole)* | $y = -0.424\ (\pm 0.22) + 1.181\ (\pm 0.05) \times x$ | 0.9927 |

($\pm 1.05$) + 15.84 ($\pm 0.25$) $\times x$, a confidence interval of 0.11 to 0.88, and $r = 0.9991$ and $r^2 = 0.9982$ are obtained. The corresponding analysis using the Raman spectra give $r$ less than 0.9700, due to the low intensity of the band at 1756 cm$^{-1}$. Comparing the IR and Raman data it can be concluded that for the systems studied the IR data showed minor improvement versus the Raman data.

The validity of these equations has been confirmed by means of the Raman spectra of pure compounds from which the ratios $I^i_{800}/I^i_{721}$ are 0.552 and 0.211, respectively. Taking into account that the intensity measured in the Raman spectrum depends on the intensity of the laser and on the factor $K_v$ including the frequency depending terms, $I_v = I_o \times K_v \times X$, and the $I^i_{800}/I^i_{721}$ ratios have been calculated. The values obtained of $0.55 \pm 1.1 \times 10^{-2}$ and $0.22 \pm 1.0 \times 10^{-2}$ are close to the experimental ones. The limit of detection (LOD) was estimated at about 0.012 mole fraction for the IR and Raman methods. Better results were obtained with IR spectra compared to the Raman spectra because, in the former, bands are characterized by higher integral absorbencies suitable for quantitative analysis.

The correlations between the results for samples with different amounts of the systems studied obtained by spectroscopic and HPLC-ESI-MS/MS techniques, demonstrate good agreement with correlation coefficients >0.9998.

The application of this mathematical model to real commercial products has been demonstrated on 10 powder samples of Cephamandole and Cefalotin containing corresponding cephamandole quantities of 1.108 g and 2.216 g, respectively. In the samples were 1.055 g of cephalotin. Three replicates were made for each sample. The IR measurements gave standard deviations of 0.011, 0.010, and 0.013 with $p$ at about 0.0500 for the systems. For the Raman data the values were 0.013, 0.011, and 0.012 with $p$ values at about 0.0550. The confidence level of >99.98% was obtained using the vibrational model presented in this chapter.

Quantitative determination of four antibiotics—cephalexin, cephalotin, cephaloglycin, and cephamandole—containing the cephalosporin nucleus in six solid binary mixtures was performed by IR and Raman spectroscopy. The $r^2$ values were within 99.32% to 99.88% and 99.90% to 95.54% for the systems studied using the absorption ratio of the bands at about 800 cm$^{-1}$ and 720 cm$^{-1}$. The absorption bands used for quantitative analysis were chosen by the detailed linear-polarized IR and Raman spectroscopy of oriented colloids in a suspension in a nematic liquid crystal. The mathematical model and correlations have been applied to 10 tablets of Cefamandole and Cefalotin giving a confidence level of >99.98 % using the vibrational model.

IR-LD spectroscopy has also been applied for the quantitative determination of three binary solid mixtures with the penicillin drugs, which represents the first such report in the literature. The obtained correlations were used for the determination of two powder samples of Amoxil® (Pfizer) and Benpen® (CSL Limited).

We also present a quantitative determination in solid binary mixtures of three penicillin antibiotics: benzylpenicillin, bacampicillin, and amoxicillin. The method appears to be unique, contrasting with the known vibrational approaches in solution, where the Raman method has been used. In the solid state, some penicillin antibiotics give a broad fluorescence band, which overlaps the Raman peaks and limits its application for quantitative analysis in solid state. Using our approach, $r^2$ values between 99.98% and 99.82% were obtained for the systems studied using the absorption ratios of the very intensive and nonoverlapped bands at 1398 cm$^{-1}$, 1685 cm$^{-1}$, and 1585 cm$^{-1}$, respectively. Another advantage is that these bands usually have low intensity in the corresponding Raman spectra, like previously reported data for cephalosporin antibiotics. The absorption bands used in the quantitative analysis were chosen by detailed linear-polarized IR spectroscopy of oriented colloids in a suspension in a nematic liquid crystal, thus giving proof of the origin of the interpreted maxima. The mathematical model used was tested on the commercial products Amoxil and Benpen giving a confidence level of >99.98$_3$%. As an independent method for quantitative determination of the studied mixtures the hybrid HPLC-ESI-MS/MS technique was used, giving in all cases good agreement with correlation coefficients >0.9998$_3$. The applicability of the method in these cases is important because for some of the pharmaceutical products it appears to be a unique vibrational tool owing to the fact that the Raman spectra can contain a broad fluorescence peak, like that of amoxicillin.

The possibilities of polarized vibrational spectroscopy of oriented colloids in a nematic host for experimental vibrational assignment, structural, and local structural elucidation have been demonstrated on N-acetyl-L-cysteine, L-cysteine, L-cystine, and L-ascorbic acid in the solid state. Supramolecular stereo structural information for all of the systems studied was compared with known crystallographic data. The method was demonstrated on the systems with the monoclinic space group P2$_1$ (Z = 2) (N-acetyl-L-cysteine, L-cysteine and L-ascorbic acid) as well as for the hexagonal P6$_1$22 space group (Z = 6) of L-cystine, which is rare for organic molecules [381]. Experimental vibrational assignment was performed for these systems. For the first time the orientation of the multicomponent solid mixture was demonstrated using the trade product ACC® (Hexal, Germany), containing N-acetyl-L-cysteine and L-ascorbic acid as main components. A quantitative IR spectroscopic approach for determination of the first compound in solid mixtures was presented. The intensity ratio of the 1716 cm$^{-1}$ peak, characteristic of N-acetyl-L-cysteine and that at 990 cm$^{-1}$, attributed to both compounds, was used. Linear regression analysis between the content and the peak ratio data for the solid-binary mixtures, gave straight-line plots: $y = 1.08_2 (\pm 0.04_9) + (-0.11_4 \pm 0.01_1) \cdot x$, where $x = 1/X_i$ with an $r$ factor of 0.9641 and a reliability value of 98.85%. The analysis by the commercial ACC 100 showed that the IR measurements gave a standard deviation of 0.010 and 0.011 with $p$ at about 0.0500 for the systems. A confidence level of greater than >98.77$_1$% was obtained.

Another example is the solid-state elucidation of the sodium salt metamizol monohydrate (Figure 5.13). In our examination, we used the trade product Analgin® (Sopharma, Bulgaria). The compound crystallizes in the monoclinic space group P2$_1$ and the unit cell contains Z = 8, oriented in the way depicted in Figure 5.14. The resulting transition moments of the in-plane and out-of-plane vibrations of the benzene ring in the framework of the whole ensemble of molecules included in the unit cell result in the elimination of the corresponding IR bands at equal dichroic ratios (Figure 5.15). This phenomenon, which cannot be observed in the isolated molecule because the transition moments of the B$_1$

and $A_1$ modes in the benzene systems are mutually orthogonally disposed, is observed in systems whose unit cell contains neighboring molecules where the disposition of the aromatic fragment leads to colinearity of the discussed modes. In the case of Analgin, the planes of the benzene rings in the neighboring molecules are inclined at an angle of 82.2(4)° (Figure 5.14).

H3C–N
N
O
H3C
N
SO3 Na H2O
H3C

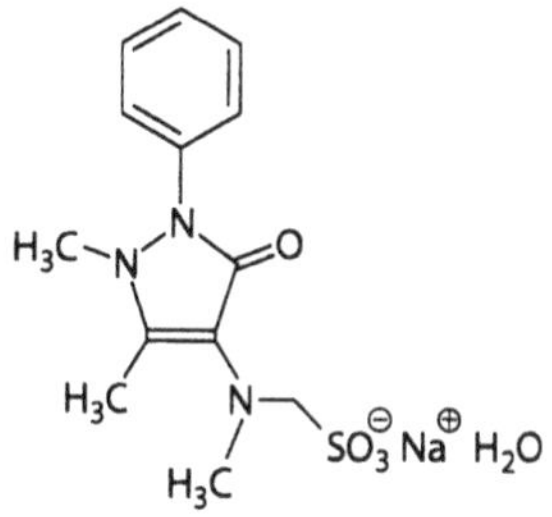

**FIGURE 5.13** Chemical diagram of the sodium salt of metamizol monohydrate (Analgin).

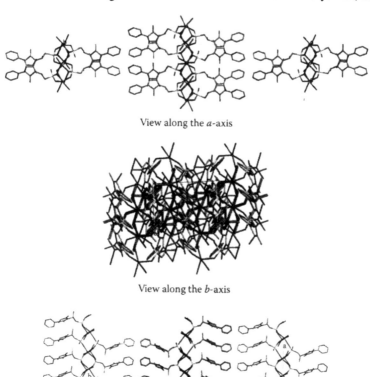

View along the *a*-axis
View along the *b*-axis
b
a
c
View along the *c*-axis

**FIGURE 5.14** Crystal structure of the sodium salt of metamizol monohydrate. (Krishna, H., Vijayan, M., and Brehm., L., 1979, *Acta Crystallogr.*, 35B: 612–613. With permission.)

The performed group analysis and IR-LD spectroscopic characterization (Figure 5.15) indicates an average orientation of the sample leading to a series of positive and negative multicomponent IR bands for which identification is limited. Additional difficulties were caused by the mutually perpendicular orientation of the molecules in the unit cell, shown in Figure 5.16.

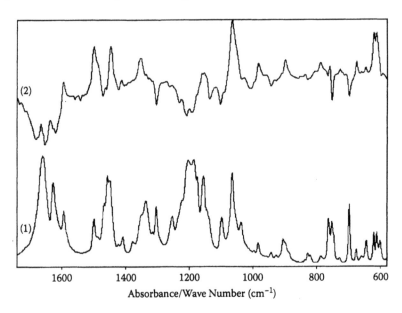

**FIGURE 5.15** (1) Nonpolarized IR and (2) difference IR-LD spectra of the sodium salt of metamizol monohydrate.

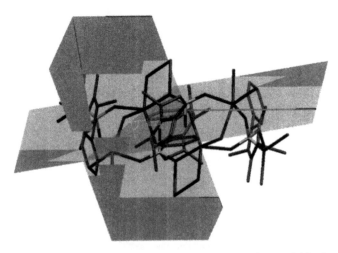

**FIGURE 5.16** Angle between the planes of the benzene rings of two neighboring molecules in the structure of the sodium salt of metamizol monohydrate. (Krishna, H., Vijayan, M., and Brehm, L., 1979, *Acta Crystallogr.*, 35B: 612–613. With permission.)

## 5.1 ANALYSIS OF MORPHINE ALKALOIDS

The polarized IR-LD spectroscopy of oriented colloids has been successfully applied for the elucidation of morphine alkaloids. Series of compounds have been characterized structurally, giving an excellent basis for future quantitative determination of these compounds. Some barbituric and thiobarbituric derivatives have also been studied (Figure 5.17 and Figure 5.18).

According to X-ray data, the 6-O-acetylcodeine molecules in a 3D structure by weak intermolecular C—H···N, C—H···π-interactions and van der Waals forces. The molecule is characterized with the classical T-shape for opiates and a dihedral angle between the mean planes of the A/B/C and D/E rings (Figure 5.17 and Figure 5.19) of 80.56(8)°. The rings are denoted following the commonly used nomenclature for opiates. The main structural features of the molecule are very similar to those of codeine, heroin, and morphine. The ring fusion and conformations are comparable to those previously reported for morphine derivatives. Aromatic ring A is planar, B

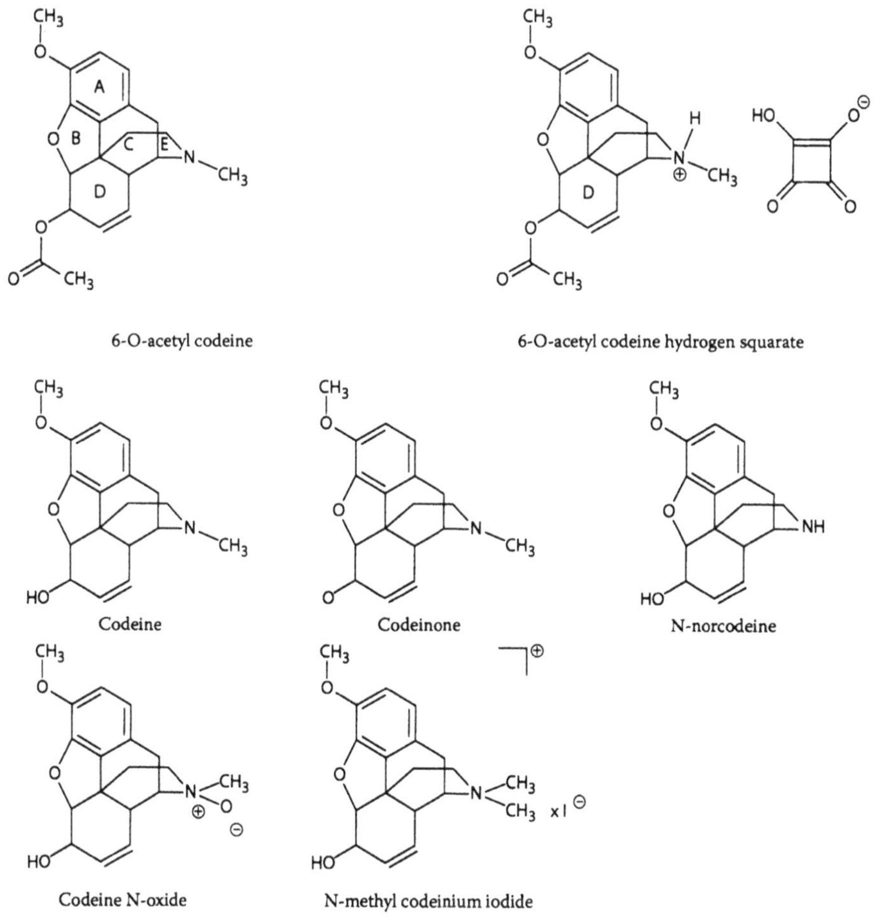

6-O-acetyl codeine

6-O-acetyl codeine hydrogen squarate

Codeine

Codeinone

N-norcodeine

Codeine N-oxide

N-methyl codeinium iodide

**FIGURE 5.17** Chemical diagram of derivatives of codeine and codeinone alkaloids.

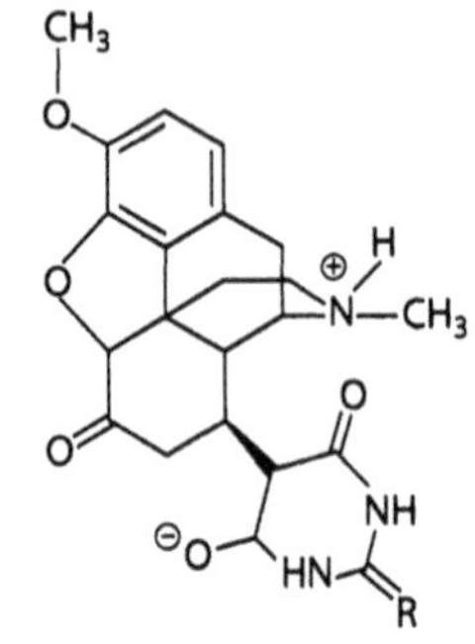

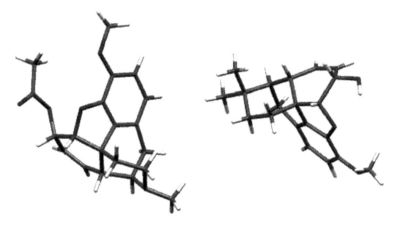

R = O or S

**FIGURE 5.18** Chemical diagram of corresponding barbituric (R=O) and thiobarbituric (R=S) derivatives.

**FIGURE 5.19** Crystal structures of 6-O-acetylcodeine and N-methylcodeinium iodide.

is close to an envelope, C and D assume half-chair conformations, and E has a chair form. Rings A and B effectively lie in one plane with a maximal deviation of B of 18.1(7)°. D and E are mutually oriented in an approximately coplanar manner and disposed at an angle of 80.56(8)°.

N-Methylcodeinium iodide crystallizes in the noncentrosymmetric space group $P2_12_12_1$. The codeine cations and iodide anions are joined by moderate intermolecular OH···I⁻ interaction (3.442 Å) (Figure 5.19). The discrete cations are disposed in a manner leading to a significant π-stacking effect with a distance of 2.980 Å. The dihedral angle between the main planes of the A/B/C and D/E rings of 86.4(5)° indicates a T-shape opiate. The main structural features of the molecule are very similar to those of codeine, heroin, and morphine. Aromatic ring A is planar, B is similar to an envelope, C and D assume half-chair conformations, and E has a chair form. Rings A and B lies in one plane with a maximal deviation of B of 3.6(3)°. D and E are mutually oriented in an approximately coplanar manner.

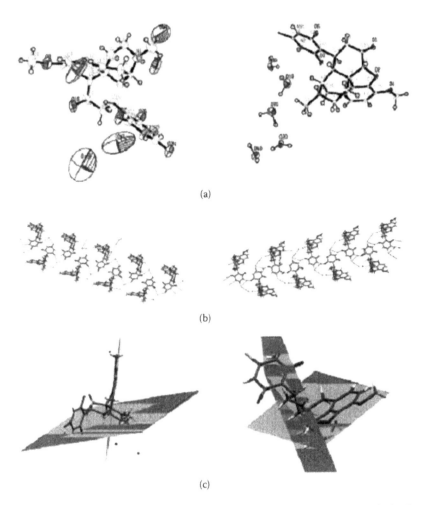

(a)

(b)

(c)

**FIGURE 5.20**   (a) The molecular structures of the barbituric and thiobarbituric derivatives with their displacement ellipsoids drawn at the 50% probability level; (b) linkage of the molecules; and (c) the dihedral angles between the mean planes of the A/B/C and D/E rings.

The barbituric and thiobarbituric derivatives depicted in Figure 5.18 also crystallize in the noncentrosymmetric space group $P2_12_12_1$ (Figure 5.20). The substituted codeinones and five solvent water molecules are joined by moderate intermolecular interactions (Figure 5.20) into infinite chains through interactions of type NH$\cdots$O (2.764 and 2.895 A, and 2.810 and 2.810 Å, respectively). The chains are stabilized in addition by interactions with the solvent water by means of hydrogen bonds of type HOH$\cdots$O (2.710 Å) and HOH$\cdots$S (3.223 Å). The protonated NH$^+$ fragment participates in an interaction (N$^+$H$\cdots$OH$_2$) with the water as well (2.809 Å and 3.223, 3.257 Å). The barbituric and thiobarbituric fragments in both compounds are flat with a deviation of 0.01° and 0.02(1)°, respectively. The C=O and C=S bonds are strikingly different as expected, with a value of 1.279 Å in the the first case and 1.704 Å in the second. The molecules exhibit the classical T-shape for opiates with a dihedral

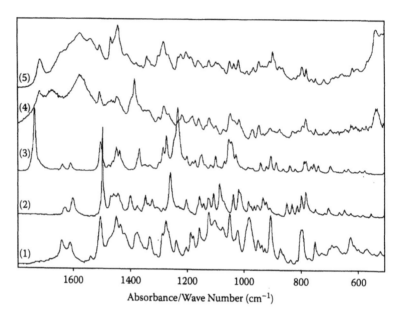

**FIGURE 5.21** IR spectra of (1) codeine N-oxide, (2) N-methylcodeinium iodide, (3) 6-O-acetylcodeine, (4) barbituric, and (5) thiobarbituric derivatives.

angle between the mean planes of the A/B/C and D/E rings of 73.0(1)° and 81.0(4)°, respectively (Figure 5.20). The rings are denoted in accordance with the commonly used nomenclature for opiates. Aromatic ring A is planar, B is close to an envelope, C and D assume half-chair conformations, and ring E has a chair form. Rings A and B are effectively coplanar with a maximal deviation for B of only 2.2(1)° and 3.1(4)°. D and E are also mutually oriented in an approximately coplanar manner.

The solid-state IR spectrum of 6-O-acetylcodeine (Figure 5.21) shows an intensive peak at 1731 cm$^{-1}$ corresponding to the $\nu_{C=O}$ stretching mode. Similar to other codeine derivatives the maxima at 1635 cm$^{-1}$, 1610 cm$^{-1}$, and 1502 cm$^{-1}$ belong to $\nu_{C=C}$, **8a** and **19a** i.p. vibrations of 1,2,3,4-o-tetrasubstituted benzenes. The adopted nomenclature (**8a**, **19a**) is according to the Wilson notation. Peaks at 1376 cm$^{-1}$ and 1367 cm$^{-1}$ can be attributed to bending modes of CH$_3$(O) and CH$_3$(N) groups ($\delta^s_{CH3(O)}$ and $\delta^s_{(CH3)N}$). The values represent a relative low-frequency shift in contrast to typical ones. The assignment of these maxima in these cases was made according to a comparison between the IR spectra of different codeine derivatives. The out-of-plane benzene 17a and 17b modes are at 906 cm$^{-1}$ and 788 cm$^{-1}$, respectively.

Simultaneous elimination of $\nu_{C=O}$, **8a** and **19a** peaks at 1735 cm$^{-1}$, 1610 cm$^{-1}$, and 1500 cm$^{-1}$ indicates the colinear disposition of the corresponding transition moments. This result correlates with single-crystal X-ray data, indicating an orientation caused by a colinearity of the $\nu_{C=O}$ and A$_1$ transition moments. On the other hand, the o.p. peaks of the last structural fragments should be also colinearly disposed due to the vanishing of B$_1$ peaks at 906 cm$^{-1}$, 788 cm$^{-1}$ and the $\gamma_{C=O}$ peak at 474 cm$^{-1}$ at the same dichroic ratio. The latter procedure does not affect the maximum at 944 cm$^{-1}$ assigned to $\gamma_{=CH}$ because it is perpendicularly oriented toward the benzene ring.

In contrast to the IR spectrum of 6-O-acetylcodeine, its hydrogensquarate is characterized with a new broad maximum in the 3200 to 2200 cm$^{-1}$ region assigned to $\nu_{NH+}$ and $\nu_{OH}$ stretching modes of the NH$^+$ group and an intermolecularly interacting OH group of the hydrogensquarate anion. In the 1800 to 1450 cm$^{-1}$ range, in parallel with the 6-O-acetylcodeine $\nu_{C=O}$, $\nu_{C=C}$, **8a** and **19a** peaks at 1735 cm$^{-1}$, 1637 cm$^{-1}$, 1610 cm$^{-1}$, 1505 cm$^{-1}$, new ones at 1804 cm$^{-1}$, 1689 cm$^{-1}$ and 1564 cm$^{-1}$ are observed. The latter bands are assigned to pairs of $\nu_{C=O}$ and $\nu_{C=C}$ and mixed $\nu_{C=O} + \nu_{C=C}$ vibrations of the hydrogensquarate anion. The IR spectrum of pure squaric acid shows the series of peaks at 1823 cm$^{-1}$, 1813 cm$^{-1}$, and 1560 cm$^{-1}$, respectively.

Similar to pure 6-O-acetylcodeine, the elimination of the 1735 cm$^{-1}$ maximum leads to disappearance of the peaks for **8a** and **19a**. In this case, the bands at 1804 cm$^{-1}$ and 1689 cm$^{-1}$ (Figure 5.21) are also eliminated. Reduction of the 790 cm$^{-1}$ peak leads to a vanishing of peak at 925 cm$^{-1}$ and the observation of a 941 cm$^{-1}$ peak, assigning their character as 17b, 17a, and $\gamma_{=CH}$ modes, respectively.

The IR spectra of compounds studied in solution show the following data:

1. Codeine, in contrast to codeinone, is characterized with a 3603 cm$^{-1}$ peak, corresponding to nonhydrogen bonded stretching vibration of the OH group ($\nu_{OH}$). N-norcodeine absorbs at 3570 cm$^{-1}$ and 3455 cm$^{-1}$ peaks, belonging to $\nu_{OH}$ and $\nu_{NH}$ modes.
2. The 3050 to 2750 cm$^{-1}$ region is typical for stretching modes of the CH$_3$, CH$_2$, and CH groups. All compounds absorb at about the 2840 cm$^{-1}$, but only codeine and codeinone possess a second peak at 2807 cm$^{-1}$. This result could assign the first maximum to $\nu^s_{CH3(O)}$ and the second one to $\nu^s_{CH3(N)}$. The obtained values are in accordance with the results in the literature for methoxy and N-methyl substituted compounds.
3. In the 1750 to 1350 cm$^{-1}$ region an intensive peak at 1677 cm$^{-1}$ in the IR spectrum of codeinone is established, belonging to the $\nu_{C=O}$ mode of the conjugated C=O group. The absence of the 1374 cm$^{-1}$ maximum indicated its character as a $\delta_{sCH3(N)}$ bending vibration and the 1388 cm$^{-1}$ peak to $\delta^s_{CH3(O)}$. Typical for all compounds, were peaks at 1633 cm$^{-1}$, 1604 cm$^{-1}$, and 1506 cm$^{-1}$ assigned as $\nu_{C=C}$, **8a** and **19a** i.p. (A$_1$) phenyl modes.
4. The 1200 to 800 cm$^{-1}$ interval showed the series of in-plane peaks of 1,2,3,4-o-tetrasubstituted benzene at 1150 cm$^{-1}$ and 1050 cm$^{-1}$, typical for all the compounds. Below 1000 cm$^{-1}$, the intensive maximum at 940 cm$^{-1}$ and a pair of maxima at 936 cm$^{-1}$ and 804 cm$^{-1}$ are observed. However, the overlapping $\gamma_{=CH}$ and **17a** and **17b** o.p. (B$_1$) modes of the aromatic fragment are observed, thus making their detailed assignment difficult by conventional IR techniques.

In contrast to IR spectra in solutions, the main differences in the solid-state spectra are established in the spectrum of N-norcodeine, where the intermolecular interactions in the condensed phase lead to low-frequency shifting of $\nu_{OH}$ and $\nu_{NH}$ to 3369 cm$^{-1}$ and 3290 cm$^{-1}$. The $\nu_{OH}$ frequency in codeine is observed at 3610 cm$^{-1}$, indicating an absence of any intermolecular interactions in solid. However, the spectrum is characterized with intensive peaks at 3540 cm$^{-1}$ and 3480 cm$^{-1}$, whose origin is difficult to assess with conventional IR spectroscopy.

The elimination of the band at 3598 cm$^{-1}$ ($\nu_{OH}$) in codeine provoked the disappearance of the 3531 cm$^{-1}$, 790 cm$^{-1}$, and 939 cm$^{-1}$ maxima, in accordance with the known crystallographic data of codeine, where the $\nu_{OH}$ transition moment is oriented close to an angle of 8° with the out-of-plane ones of the 1,2,3,4-tertasubstituted benzene ring in the frame of the unit cell. The observed maximum at 944 cm$^{-1}$, therefore, corresponds to an o.p. mode ($\gamma_{=CH}$) of the C=C group, with a transition moment near to perpendicular toward the benzene ($B_1$) ones, at an angle of 84.5°. The simultaneous elimination of the 1607 cm$^{-1}$ peak (**8a**, $A_1$), the **19a** one at 1502 cm$^{-1}$, and $\delta^s_{CH3(N)}$ at 1376 cm$^{-1}$, is also in accordance with single-crystal X-ray data indicating the colinearity of corresponding transition moments in the framework of the molecules included in the unit cell. The fact that the solid-state peaks at 3465 cm$^{-1}$ and 1656 cm$^{-1}$ vanish at the same dichroic ratio suggests that these maxima result from the crystal field effect.

The RDP applied to solid-state IR-LD spectra of codeinone resulted in the elimination of $\nu_{C=O}$ at 1670 cm$^{-1}$, a disappearance of the 1636 cm$^{-1}$ ($\nu_{C=C}$) peak, and a strong reduction of 8a, 19a, and $\delta^s_{CH3(N)}$ peaks at 1610 cm$^{-1}$, 1502 cm$^{-1}$ and 1372 cm$^{-1}$, indicating the colinear disposition of their transition moments. The elimination of the 800 cm$^{-1}$ peak at the same dichroic ratio as the 936 cm$^{-1}$ peak indicates their **17a** and **17b** mode origins. The observed peak at 943 cm$^{-1}$ corresponds to $\gamma_{=CH}$, typical for a cis-1,2-disubstituted C=C bond.

The reduction of $\nu_{NH}$ at 3295 cm$^{-1}$ in the polarized IR spectra of N-norcodeine leads to the practical elimination of the $\nu_{OH}$ peak, while the application of the same procedure on 8a at 1602 cm$^{-1}$ leads to the disappearance of 19a at 1506 cm$^{-1}$ and the peaks at 1133 cm$^{-1}$ and 1052 cm$^{-1}$, belonging to the 9a and 18a in-plane modes. A simultaneous reduction of the peaks at 798 cm$^{-1}$ and 944 cm$^{-1}$ occurs, thus assigning their o.p. character of the substituted phenyl ring. The occurrences during the latter procedure indicate that the maximum at 954 cm$^{-1}$ correspond to $\gamma_{=CH}$.

The solid-state IR spectrum of N-norcodeine ester amide of squaric acid ethyl ester is characterized by a complicated spectral curve, for which detailed IR characteristic band assignment requires a preliminary deconvolution followed by spectral CFP. A broad maximum in the whole 3000 to 2000 cm$^{-1}$ region corresponds to the $\nu_{OH}$ stretching mode where an intermolecular interaction with the OH group in N-norcodeine ester amide is observed. In addition, a broad band in the 1700 to 700 cm$^{-1}$ IR spectral region supported an OH$\cdots$O=(C) bond formation as well as the low-frequency shift of the $\nu_{CO}$ peak to 1792 cm$^{-1}$ in the squarate fragment. In the cases of noninteracted squarate derivatives, the discussed peak is observed at above 1810 cm$^{-1}$. The CFP applied on the IR spectrum leads to an observation of a series of peaks with the following assignments: 1792 cm$^{-1}$ ($\nu^s_{CO}$), 1688 cm$^{-1}$ ($\nu^{as}_{CO}$), and 1560 cm$^{-1}$ ($\nu_{C=C}$) of the squarate part; and 1646 cm$^{-1}$ ($\nu_{C=C}$), 1597 cm$^{-1}$ (**8a**), 1520 cm$^{-1}$ (**8b**), and 1503 cm$^{-1}$ (**19a**). The last three maxima correspond to radial vibrations of the 1,2,3,4-o-tetrasubstituted aromatic fragment.

Simultaneous elimination of the 940 cm$^{-1}$ and 784 cm$^{-1}$ peaks occurs, that is typical for o.p. **17b** and **17a** ($B_1$) modes of the aromatic fragment in the molecule, and a 955 cm$^{-1}$ ($\gamma_{=CH}$ out-of-plane mode) peak is established. In parallel, the 1792 cm$^{-1}$ peak vanishes.

The IR spectrum of codeine-N-oxide is characterized with an intensive doublet of bands at 3402/3199 cm$^{-1}$ belonging to FR splitting absorption maxima for the stretching vibration ($v_{(OH)}$) of the OH group of codeine. Direct evidence of this assumption is provided by the elimination of these submaxima at equal dichroic ratios. The profile of the spectroscopic pattern and the low-frequency shifting suggest participation of the OH group in the intermolecular interaction of OH···O type. Additional evidence for this conclusion is seen in the multicomponent band within the 2600 to 2300 cm$^{-1}$ range, typical for the hydrogen bonded systems like acids where the effect is explained by FD resonance. Moreover, these bands possess different polarization, typical for BS. Detailed explanations of the phenomenon can be found. The region 3080 to 2800 cm$^{-1}$ (typical for stretching modes of CH$_3$, CH$_2$, and CH groups) is characterized with the absorption band at 2843 cm$^{-1}$ and 2815 cm$^{-1}$, assigned to $v^s_{CH3(O)}$ and $v^s_{CH3(N)}$, stretching vibrations, respectively. The latter band is shifted to lower frequency, compared with the data for other codeines, where the discussed vibration is observed between 2825 and 2835 cm$^{-1}$. The (C)(O)N(C)$_2$ fragment can be described with local distorted pyramidal symmetry and the bands at 1376 cm$^{-1}$ correspond to the $\delta^s_{CH3(N)}$ bending vibration. The band at 1385 cm$^{-1}$ can be assigned to the $\delta^s_{CH3(O)}$ mode. The $v_{C=C}$ and **8a** vibrations are observed at 16112 cm$^{-1}$. Typical for other derivatives, IR bands at 1500 cm$^{-1}$ (**19a** i.p. (A$_1$) *Ph* modes) are of low intensity and practically undefined. To $v_{C=C}$ can be assigned the maximum at 1638 cm$^{-1}$. In the 1200 to 800 cm$^{-1}$ region, a series of in-plane peaks of 1,2,3,4-o-tetrasubstituted benzene at 1150 cm$^{-1}$ and 1050 cm$^{-1}$ are observed. Below 1000 cm$^{-1}$, intensive maxima at 930 to 921 cm$^{-1}$ and 799 to 784 cm$^{-1}$ are observed. They can be assigned to $\gamma_{=CH}$ of the double CH=CH bond as well as 17a, 17b, and 11-$\gamma_{CH}$ o.p. modes of aromatic fragment. Direct evidence for this assumption is seen by the elimination of these maxima at equal dichroic ratios in corresponding polarized IR-LD spectra. The reduction leads to an appearance of the maxima for differently disposed molecules in the unit cell. This phenomenon can be illustrated using the reduced IR spectrum, where the elimination of the band of the o.p. bending vibration at 752 cm$^{-1}$ in the inflex point thus additionally confirms the BS effect. Codeine-N-methyl iodide is also an example of this phenomenon. The compound crystallizes in the orthorhombic system with the P2$_1$2$_1$2$_1$ space group.

Evidence of the stabilization of the zwiterionic structure of barbituric and thiobarbituric derivatives in the solid state is obtained by looking at the IR spectra of both compounds in the solid state. The IR spectroscopic region within the whole 3500 to 2500 cm$^{-1}$ region is characterized with a broad, strongly overlapped multicomponent band, typical for the amino acids and small peptides, and assigned to the $v_{N+H}$ stretching vibration. The FR, observed in the previously discussed systems, leads to overlapping of the IR characteristic bands within the 3050 to 2750 cm$^{-1}$ region, that are typical for stretching modes of the CH$_3$, CH$_2$, and CH groups. The intensive bands at 3600 cm$^{-1}$ and 3450 cm$^{-1}$ belong to stretching $v_{OH}$ vibrations of the solvent molecules. The bands at about 1714 cm$^{-1}$ correspond to stretching $v_{C=O}$ vibration of the codeine fragment. The in-plane vibrations of the aromatic fragment are observed at 1575 cm$^{-1}$ and 1507 cm$^{-1}$ and are weakly influenced by the type of barbituric substituent. The maxima are eliminated at equal dichroic ratios in pairs depending on their $A_1$ or $B_2$ local symmetry. More significant are the differences of the IR-spectroscopic patterns of both the compounds associated with the bending $\delta_{N+H}$ and $\delta_{NH}$ vibrations. According to the crystallographic data, the intermolecular interactions of the discussed fragments differ, and for this reason the bending vibrations are observed at higher frequencies in the barbituric derivative (1675 cm$^{-1}$ and 1545 cm$^{-1}$) as compared to the frequencies of the thiobarbituric derivative (1640 cm$^{-1}$ and 1535 cm$^{-1}$).

# 6 Application in the Chemistry of Dyes

## 6.1 STILBAZOLIUM SALTS

Interest in stilbazolium salts during the past 30 years has been due to many of the derivatives possessing large second-order molecular hyperpolarizability. The second-harmonic generation (SHG) properties of these dyes have attracted much attention because of their application in nonlinear optics. Their second-order non-linear optical (NLO) properties are very sensitive to the symmetry of corresponding crystals [384]. The variation of the type of the counterion in the salts provides a simple and highly successful strategy to generate materials with larger $\chi$ values. This methodology has been also supported by the observed crystal structure and properties of DAST [384–386]. 4-Dimethylamino-N-methyl-4-stilbazolium tosylate (DAST) is well known as a novel terahertz (THz)-wave emitter as well. Systematic studies of the NLO properties of these dyes have been presented [387–428]. Stilbazolium salt assemblies in the condensed phase offer important functionality in nature and technology. The aggregation processes have been intensively studied [429–433].

The possibilities of linear-dichroic infrared (IR-LD) spectroscopy of oriented colloids in nematic liquid crystals (NLC) on stilbazolium salts have been presented [213–218,434,435]. Substituted dyes (Figure 6.1) have been investigated, illustrating the advantages of polarized IR and Raman spectroscopy in the solid state when compared to the "classical" methods of ultraviolet-visible spectroscopy (UV-VIS) and fluorescence spectroscopy for the analysis of dyes. The focus is on obtaining additional information on the structure and spectroscopic properties for these systems in the solid-state. Moreover, it is well known that the dyes have difficulty crystallizing and their characterization by the method of single-crystal X-ray diffraction is limited. Crystallographic studies of, for example, 4-OH substituted stilbazolium salts are rare (Figure 6.1 and Table 6.1).

This was the motive for the systematic structural elucidation of these derivatives using the possibilities of IR-LD spectroscopy of oriented colloids. The method allows structural and local structural characterization as well as experimental assignment of the IR bands. We are also able to describe the corresponding vibrations in the Raman spectra of the stilbazolium salts. In some of the examples included in the book, we consider the application of the IR-LD and Raman spectroscopy for analysis of the crystals of dyes, where the so-called crystallographic disorder is often observed. The key role of the OH group in the 4′-position (Figure 6.1) is associated with the possibility for obtaining the different tautomers, depending on the pH value of the medium.

The possible redistribution of the electronic density in these compounds as typical push–pull systems depends on the solvent polarity. When the electron transition

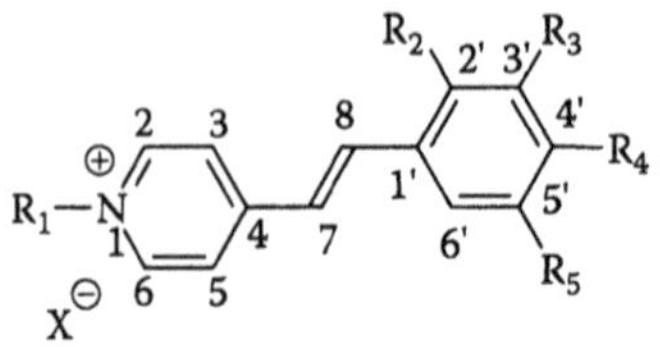

| $R_1$ | $R_2$ | $R_3$ | $R_4$ | $R_5$ | $X^-$ |
|---|---|---|---|---|---|
| $CH_2CH_2CH_3$ | H | $OCH_3$ | OH | $OCH_3$ | Cl |
| $CH_2(CH_2)_2CH_3$ | H | $OCH_3$ | OH | $OCH_3$ | |
| $CH_3$ | H | $OCH_3$ | OH | $OCH_3$ | |
| $CH_3$ | H | H | OH | $OCH_3$ | 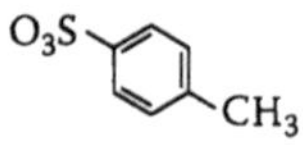 $O_3S$- (tolyl) |
| $CH_3$ | H | H | $N(CH_3)_2$ | H | |
| H | H | H | $N(CH_3)_2$ | H | S(+) tartarate |
| $CH_3$ | H | H | OH | H | (squarate) |
| H | H | H | $N(CH_3)_2$ | H | |
| $CH_3$ | H | H | OH | H | |
| $CH_3$ | H | H | OH | H | $H_2PO_4$ |
| $CH_3$ | H | H | OH | H | I |
| $CH_3$ | OH | H | H | H | |
| $CH_3$ | H | OH | H | H | |
| $CH_2(CH_2)_8CH_3$ | OH | H | H | H | |
| $CH_2(CH_2)_8CH_3$ | H | OH | H | H | |
| $CH_3$ | OH | H | $OCH_3$ | H | |
| $CH_3$ | OH | H | H | $OCH_3$ | |
| $CH_2(CH_2)_8CH_3$ | OH | H | H | $OCH_3$ | |

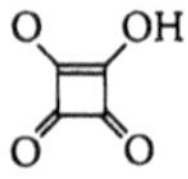

**FIGURE 6.1** Chemical diagram of substituted stilbazolium salts.

is connected with intramolecular charge transfer (CT), this leads to a significant difference between the dipole moment in the ground and excited states that determines their significant solvatochromic effect or NLO properties in solution. Depending on the solvent polarity of the CT band in 1-methyl-4-[2-(4-hydroxyphenyl)ethenyl)] pyridinium salts, different types of counterions (Figure 6.1) exhibit a bathochromic

## TABLE 6.1
## Crystal Structures of 4-OH Substituted Stilbazolium Salts

| $R_1$ | $R_3$ | $R_5$ | X- | CCDC Code[a] |
|---|---|---|---|---|
| CH₃ | H | H | (+)-Camphor-10-sulfonate | BOJWAY |
| CH₃ | OH | H | Toluene-p-sulfonate | JOYBEE10 |
| CH₃ | OCH₃ | H | Bromobenzenesulfonate | LANLAO |
| CH₃ | H | H | Triiodide iodide | LIBZEC |
| 2-Hydroxyethyl | | H | 2,4-Dihydroxyacetophenone | NUPFUZ[b] |
| Butyl | OH | H | Cl- | OMIXAJ |
| CH₃ | OCH₃ | H | Ptoluenesulfonate | OMIXUD, VIJPAF |
| CH₃ | OCH₃ | H | Chlorobenzenesulfonate | WAQNOS |
| CH₃ | OCH₃ | | I- | QAWMIK |
| 2-Hydroxyethyl | H | | I- | QIGHET |
| CH₃ | H | | | SIFFAO[b] |
| 2-Hydroxyethy | H | | | SUJKUD[b] |
| CH₃ | H | | Tetraphenylborate dimethyl, sulfoxide solvat | TAQQEI |
| CH₃ | H | | 2-Amino-4-nitrophenol | TEBLUH, TEBMAO |
| 2-Hydroxyethy | H | | 2,4-Dihydroxybenzaldehyde | TETQAK |
| CH₃ | OCH₃ | | I- | TOHGIG |
| CH₃ | H | | Bromide iodide | YIFPIN |

*Source:* Kolev, T.S., B.B. Koleva, M. Spiteller, H. Mayer-Figge, and W.S. Sheldrick, 2008, *Dyes Pigm.*, 79:7–13. (With permission.)

[a] CCDC—Cambridge Crystallographic Database Centre, http://www.ccdc.cam.ac.uk/.
[b] Quinoide form of the dyes obtained after deprotonation of the 4-OH group.

shift of up to 70 nm ongoing from $H_2O$ to 1,2-dichloromethane or cyclohexane. The obtained negative solvatochromic effect in these compounds has been explained by intra- and intermolecular charge transfer. A classical example is 1-ethyl-4-methoxy-carbonylpyridinium iodide, whose negative sovatochromism, has been explained with that in the ground state. It forms ion pairs and the CT band corresponds to the intermolecular charge transfer from the iodide ion to the pyridinium cation. In the ground state the dipole moment is significant, contrary to the excited state, where $\mu_e$ have a lower value due to radical formation. In solvent mixtures at a molar ratio 1:1, two or more bands are often registered. This phenomenon is also often observed in acetonitrile. The observation of these maxima as well as the spectral changes and the well-defined isobestic points provide good evidence for an equilibrium between monomeric and dimeric species. It is noteworthy that in addition to the very intense hypsochromically shifted absorption band for the H-dimer, a weak band appears at a longer wavelength, which can be ascribed to the forbidden transition to the lower energy exciton state [431,436–442].

In the case of 1-methyl-4-[2-(4-hydroxyphenyl)ethenyl)]pyridinium] hydrogen-phosphate in an ethanol:water solvent mixture (1:1), two bands at 386 nm and 492 nm are observed. The observation of two maxima in acetonitrile as well as the spectral changes and the well-defined isosbestic point at 450 nm provide good evidence for an equilibrium between monomeric and dimeric species. It is noteworthy that in addition to the very intense hypsochromically shifted absorption band for the H-dimer, a weak band appears at a longer wavelength, which can be ascribed to the forbidden transition to the lower energy exciton state.

With respect to the development of new materials with potential NLO applications, compounds with quinoide-like forms, obtained after the deprotonation of the OH-group in the 4'-position, can be reasonably studied because a significant bathochromic effect is observed (Table 6.2). Negative solvatochromism is observed with lowering of the solvent polarity and a shifting of $\lambda_{max}$ occurs within 170 to 210 nm.

The UV-VIS spectrum of 1-methyl-4-[2-(4-oxocyclohexadienyliden)ethyliden]-1-4-dihydropyridine in acetonitrile is completely different from that of 1-methyl-4-[2-(4-hydroxyphenyl)ethenyl)]pyridinium] iodide. In the former, two bands at 388 nm and 566 nm and a well-defined isosbestic point at 455 nm are recorded. The observed characteristics in acetonitrile provide good evidence for equilibrium between monomeric and dimeric species [215]. However, the Vis-spectrum 1-methyl-4-[2-(4-hydroxyphenyl)ethenyl)]pyridinium] hydrogensquarate in ethanol is characterized with two bands at 514 nm and 621 nm. The presence of two maxima in acetonitrile again manifested the H-type of aggregation similar to other merocyanine dyes. The different position of the absorption maxima suggested the participation of hydrogensquarate anion in a different manner. These assumptions are supported with high-performance liquid chromatography–electrospray with tandem mass spectrometry, using electrospray ionization (HPLC-ESI-MS/MS) data as well. The most intensive signal in the mass spectrum is observed at $m/z$ 326.88 corresponding to the singly charged $[C_{18}H_{16}NO_5]^+$ cation with molecular weight of 326.32. The data indicate a cation–anion association in the gas phase. The next peak at 213.34 is assigned to $[C_{14}H_{15}ON]^+$ with the molecular weight of 213.22 of the dye.

Using the empirical parameter of the solvent polarity Z based on the molar energy of the transition in $E_T$, in kilocalorie per mole (kcal/mol) for the CT band, the $E_T$ values of sodium salt of 1-methyl-4-[2-(4-hydroxyphenyl)ethenyl)]pyridinium] hydrogensquarate are shown in Table 6.3. These data illustrate the analogy between the Z values of the sodium salt of 1-methyl-4-[2-(4-hydroxyphenyl)ethenyl)]pyridinium] hydrogensquarate with $E_T(30)$, which is the empirical solvent polarity parameter, based on the intramolecular charge transfer absorption of a pyridinium-N-phenolate betaine dye. The Z values are practically equal in the solvents acetone, pyridine, and cyclohexane, which means that for a difference in the values of $\varepsilon$, solvent donor number (DN) and solvent acceptor number (AN) of 35.3, 33.1, and 18.9 (kcal/mol), the

---

## TABLE 6.2
## UV-Spectra of 1-methyl-4-[2-(4-oxocyclohexadienyliden)ethyliden]-1,4-dihydropyridine

| Solvent | $H_2O$ | $CH_3OH$ | $C_2H_5OH$ | 2-Propanole | $CH_2Cl_2$ |
|---|---|---|---|---|---|
| $\lambda_{max}$ [nm] | 444 | 482 | 514 | 525 | 614 |
| log $\varepsilon$ | 3.808 | 4.430 | 4.592 | 4.543 | 3.211 |

## TABLE 6.3
### $E_T$ Values kcal/mol of the Sodium Salt of 1-methyl-4-[2-(4-hydroxyphenyl)ethenyl)]pyridinium] Hydrogensquarate at Standard $E_T(30)$

| Solvent | $E_T(30)$ [39] [kcal/mol] | Compound Studied |
|---|---|---|
| $H_2O$ | 63.1 | 64.4 |
| $CH_3OH$ | 55.4 | 59.3 |
| $C_2H_5OH$ | 51.9 | 55.6 |
| 2-propanol | 48.4 | 54.5 |
| $CH_2Cl_2$ | 39.1 | 46.6 |
| Acetone | 42.2 | 48.6 |
| $C_5H_5N$ | 40.5 | 48.0 |
| Cyclo-$C_6H_{12}$ | 30.9 | 48.3 |

## TABLE 6.4
### Difference of $\Delta\lambda_{max}$ versus $\Delta T$ in the Case of Sodium Salt of 1-methyl-4-[2-(4-hydroxyphenyl)ethenyl)]pyridinium] Hydrogensquarate

| Solvent | $\Delta\lambda_{max}$ [nm] | $\Delta T$ [°C] | Solvent | $\Delta\lambda_{max}$ [nm] | $\Delta T$ [°C] |
|---|---|---|---|---|---|
| $H_2O$ | 9 | 75 | $CH_3OH$ | 19 | 65 |
| $C_2H_5OH$ | 36 | 95 | $C_3H_7OH$ | 40 | 105 |
| $(CH_3)_2C=O$ | 40 | 65 | $CH_2Cl_2$ | 37 | 65 |

energy of the transition is kept equal. In the case of the sodium salt of 1-methyl-4-[2-(4-hydroxyphenyl)ethenyl)]pyridinium] hydrogensquarate, only solvents with a significant level of AN lead to the stabilization of a quinoide-like structure and determines the solvatochromism of 167 nm. Schematically, the phenomenon could be explained with dipole–dipole solute–solvent interactions with participation of a hydrogensquarate ion. As far as Z depends on the temperature, we obtained a dependence of $\lambda_{max}$ versus temperature for the sodium salt of 1-methyl-4-[2-(4-hydroxyphenyl) ethenyl)]pyridinium] hydrogensquarate, shown in Figure 6.4. With decreasing of the temperature, a hypsochromic shifting of the CT band is obtained (Table 6.4). The decreasing of the Z values with an increasing of the temperature shows that the solute–solvent interactions are weaker at a higher temperature.

The role of the N-aliphatic and OCH$_3$ substituents can be illustrated by looking at the UV-VIS spectroscopic data of 1-butyl-4-[2-(3,5-dimethoxy-4-hydroxyphenyl)ethenyl)]piridinium] chloride. A bathochromic shifting of $\lambda_{max}$ with 21 nm (Figure 6.2) from 1,2-dichloromethane to acetonitrile is obtained. However, the quinoide form is characterized with a solvatochromic effect of 140 nm (Figure 6.2). In addition, in acetone and acetonitrile two bands at about 600 and 640 nm and a shorter wavelength, one at 560 nm are observed.

The distribution of the highest occupied molecular orbital (HOMO) and lowest unoccupied molecular orbital (LUMO) for the ground state of 1-butyl-4-[2-(3,5-dimethoxy-4-hydroxyphenyl)ethenyl)]piridinium] cation is illustrated in Figure 6.3.

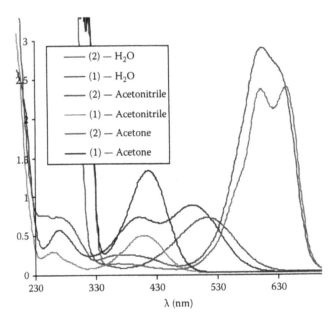

**FIGURE 6.2** UV-Vis spectra of 1-butyl-4-[2-(3,5-dimethoxy-4-hydroxyphenyl)ethenyl)] piridinium] chloride tetrahydrate in different media at a concentration of $2.5.10^{-5}$ M.

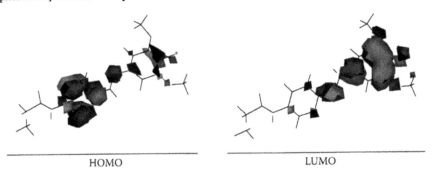

HOMO                    LUMO

**FIGURE 6.3** Molecular orbital surface of the HOMO and LUMO for the ground state of the quinoide form (2) of the 1-butyl-4-[2-(3,5-dimethoxy-4-hydroxyphenyl)ethenyl)]piridinium cation.

Nearly all of the molecular orbital (MOs) are substantially localized on the conjugated plane, with only small contributions to the group out of the plane. In contrast, the underlined CT of the quinoide form is observed (Figure 6.3).

For the precise investigation of the geometry changes associated with the electronic excitation state, of all of the derivatives depicted in Figure 6.1, the geometry of the compounds in the singlet excited state was optimized at the CIS/6-31++G** level of theory for comparison with the data for the ground state optimized at HF/6-31++G**. The results indicate that the structural shift is predominantly localized in the conjugated plane in the aromatic form and that the groups displaced from the discussed plane do not change significantly. The distribution for the HOMO and

LUMO of the lowest single excited state shows a strong optical emission in the case of the corresponding quinoide forms (Figure 6.3).

The role of the effect of the substituents on the spectroscopic characteristics of the stilbazolium salts has been elucidated in detail using the corresponding derivatives, where the OH and $OCH_3$ groups are at the 2-position and/or 3-position (Figure 6.1). The role of the N-aliphatic fragment has also been studied.

Similar to the 4-OH substituted derivatives, the pH values influenced the equilibrium between the aromatic and quinoide form (Figure 6.4) also for the corresponding 2-OH derivatives. In the case of 3-OH substituted salts, the pH values had a weaker effect on the spectroscopic characteristics.

The quinoide forms of 2-OH derivatives are characterized by large bathochromic shifts of their CT bands in comparison to the aromatic forms (Figure 6.5). The intensity of the bands is only weakly affected by the deprotonation process. The aliphatic chain at the $R_1$ position has a small bathochromic effect on the CT band of about 5 nm.

The spectroscopic data for the 3-OH derivatives also showed a bathochromic shifting of the CT band at about 260 nm, whose magnitude depends on the pH of the

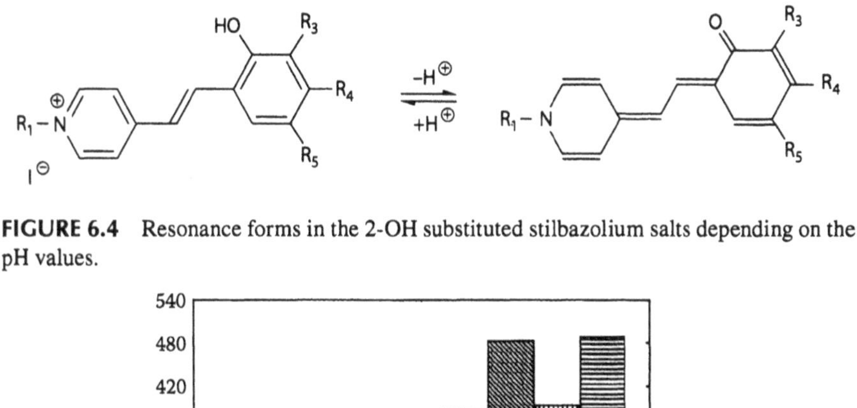

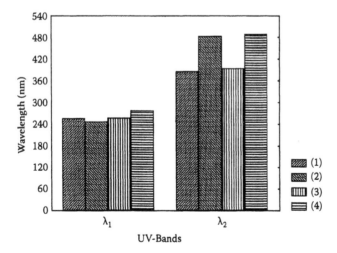

**FIGURE 6.4**  Resonance forms in the 2-OH substituted stilbazolium salts depending on the pH values.

**FIGURE 6.5**  UV-Vis spectroscopic data for the aromatic and quinoide forms of 2-OH derivatives: (1) 1-methyl-4-[2-(2-hydroxyphenyl)ethenyl]-pyridinium iodide, (2) 1-methyl-4-[2-(2-oxocyclohexadienyliden)ethylidene-1,4-dihydropyridine, (3) 1-decyl-4-[2-(2-hydroxyphenyl)ethenyl]-pyridinium iodide, and (4) 1-decyl-4-[2-(2-oxocyclohexadienylide n)ethyliden]-1,4-dihydropyridine.

medium (Figure 6.6). However, in the case of dodecyl 3-OH an additional band at 420 nm is observed at high pH values, indicating different association effects.

In the cases of the substituted 2-OH derivatives, the second OCH₃ substituent in the benzene ring leads to a stronger bathochromic effect on the CT bands of corresponding quinoide forms at higher pH values as well as a pronounced hyperchromic effect (Figure 6.7). The stronger bathochromic effect of the quinoide forms is

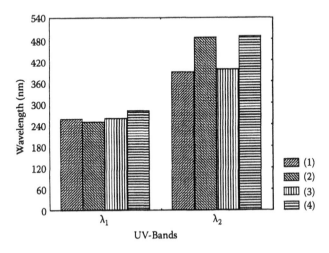

**FIGURE 6.6**   UV-Vis spectroscopic data for the aromatic and quinoide forms of 3-OH derivatives: (1) 1-methyl-4-[2-(3-hydroxyphenyl)ethenyl]-pyridinium iodide, (2) 1-methyl-4-[2-(3-oxocyclohexadienyliden)ethylidene-1,4-dihydropyridine, (3) 1-decyl-4-[2-(3-hydroxyphenyl)ethenyl]-pyridinium iodide, and (4) 1-decyl-4-[2-(3-oxocyclohexadienyliden)ethyliden]-1,4-dihydropyridine.

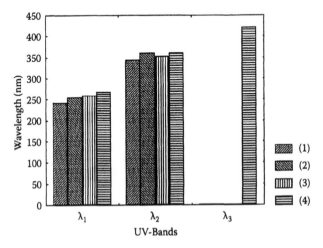

**FIGURE 6.7**   UV-Vis spectroscopic data for (1) 1-methyl-4-[2-(4-methoxy-2-hydroxyphenyl)ethenyl]-pyridinium iodide, (2) 1-methyl-4-[2-(5-methoxy-2-methoxy-2-hydroxyphenyl)ethenyl]-pyridinium iodide, (3) 1-decyl-4-[2-(4-methoxy-2-hydroxyphenyl)ethenyl]-pyridinium iodide, and (4) 1-decyl-4-[2-(5-methoxy-2-methoxy-2-hydroxyphenyl)ethenyl]-pyridinium iodide.

underlined in 2-OH derivatives, and for all the decyl-derivatives, different aggregates are observed in solution (Figure 6.8). The presence of the $OCH_3$ group on the opposite side of the OH function leads to a stronger bathochromic shift of the CT band.

As can be seen, tautomers and small aggregates can be easily studied in solution by means of the electronic spectra. Such information can also be obtained by means of ${}^1H$ and ${}^{13}C$-NMR in solution. The stabilization of quinoide-like forms in the 4-OH substituted stilbazolium salts after the deprotonation of the OH group showed a shifting of the proton signals of the aromatic protons in the deprotonated forms within the range 8.50 to 9.21 ppm (C2, C6), and to 8.00 ppm, 7.50 ppm, and 7.70 ppm (C3, C5, C7, and C8). The values of the aromatic protons signals are 7.00 ppm (H-2'), 6.80 ppm (H-3'), 7.24 ppm (H-5'), and 7.50 ppm (H-6'). The ${}^{13}C$-NMR signals of the aromatic forms are 132.6 ppm (C2, C6), 122.45 ppm (C3, C5), 152.55 ppm (C4), 118.12 ppm (C7), and 131.85 ppm (C8). The stabilization of the quinoide-like structure leads to a shifting of the proton chemical signals of C2, C6, C3, C5, C7, and C8 with Δδ of 1.24, 0.77, 0.70, and 1.00 ppm. The aromatic proton chemical signals are shifted as well and Δδ values are observed within the 0.55 to 1.03 ppm range.

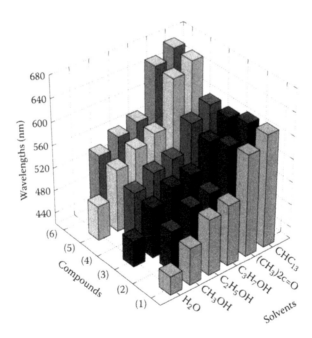

**FIGURE 6.8** Vis spectroscopic data for the aromatic and quinoide forms of 2-OH derivatives in different solvents: (1) 1-methyl-4-[2-(4-methoxy-2-hydroxyphenyl)ethenyl]-pyridinium iodide, (2) 1-methyl-4-[2-(4-methoxy-2-oxocyclohexadienyliden)ethyliden]-1,4-dihydropyridine, (3) 1-methyl-4-[2-(5-methoxy-2-hydroxyphenyl)ethenyl]-pyridinium iodide, (4) 1-methyl-4-[2-(5-methoxy-2-oxocyclohexadienyliden)ethyliden]-1,4-dihydropyridine, (5) 1-decyl-4-[2-(5-methoxy-2-hydroxyphenyl)ethenyl]-pyridinium iodide, and (6) 1-decyl-4-[2-(5-methoxy-2-oxocyclohexadienyliden)ethyliden]-1,4-dihydropyridine. **(See color insert.)**

However, the solid-state π-interactions, structural, and local structural elucidation can be obtained by linear-polarized IR and Raman spectroscopy. In our description we make a parallel between the IR-LD spectroscopic results with crystallographic data as well as with theoretical quantum chemical calculations for the electronic structure and vibrational properties of the dyes. The latter data support the experimental vibrational assignment as well as the structural results, especially in the cases where it is impossible to obtain suitable results by single-crystal X-ray diffraction.

For all of the studied dyes, a significant degree of macroorientation in the polarized IR spectrum of the salts facilitates an adequate interpretation of the polarized IR data. In all the cases the characteristic IR and Raman bands of corresponding counterions in the salts must be taken into account when studying the vibrations of the differently substituted dyes (Figure 6.9). For example, in the case of 1-methyl-4-[2-(4-hydroxyphenyl)ethenyl)]pyridinium] hydrogenphosphate the broad absorption band between 3000 and 1800 cm$^{-1}$ is a result of the stretching vibration of the OH group ($v_{OH}$) of $H_2PO_4^-$ anions and the bands at 1270 cm$^{-1}$ and 1170 cm$^{-1}$ are caused by the stretching vibrations of the P=O and P-O groups ($v_{P=O}$ and $v_{P-O}$). All of the OH-substituted salts (Figure 6.1) are characterized with a relatively strong intensive band in their solid-state IR spectra between 3500 and 3100 cm$^{-1}$, depending on the type of the intermolecular hydrogen bonding with participation of the OH group. Usually, this band is absent in the corresponding Raman spectra (Figure 6.9). More powerful in the IR-LD analysis is the appearance of the IR and Raman bands within the regions 1700 to 1450 cm$^{-1}$ and 1000 to 680 cm$^{-1}$, where the in-plane (i.p.) and out-of-plane (o.p.) vibrations of the disubstituted benzene and pyridine fragments in the stilbazolium salts are observed. Detailed assignments of these vibrations have been reported by Varsanyi [443]. The vibrational characteristics of the benzene moiety

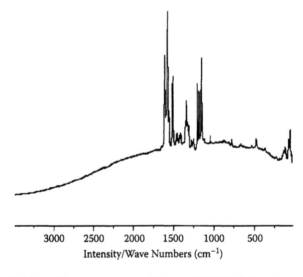

$$3000 \quad 2500 \quad 2000 \quad 1500 \quad 1000 \quad 500$$

Intensity/Wave Numbers (cm$^{-1}$)

**FIGURE 6.9**   Solid-state Raman spectrum of 1-methyl-4-[2-(3,5-dimethoxy-4-hydroxyphenyl)ethenyl)]pyridinium] chloride.

are observed at a lower frequency than those of the pyridinium fragment. The corresponding maxima of the $A_1$ symmetry class are more intensive in the Raman spectra, whereas it is the $B_1$ symmetry class in the IR spectra of the dyes. The transition moments of corresponding $A_1$ and $B_1$ vibrations are mutually perpendicular in the frame of the given structural fragment, and they are attractive for IR-LD analysis. Relatively intensive bands in both the IR and Raman spectra are given by the symmetric bending $CH_3$ vibration at 1340 cm$^{-1}$ ($\delta^s_{CH3}$) in the case of N-methyl derivatives, with a difference of $\pm 4$ cm$^{-1}$.

The IR spectrum of 1-methyl-4-[2-(4-hydroxyphenyl)ethenyl)]pyridinium] hydrogenphosphate shows a relatively intensive band at 3224 cm$^{-1}$ ($\nu_{OH}$) belonging to the hydrogen bonded OH group, thus correlating with the crystallographic data (see later). The aromatic in-plane modes of the benzene and pyridine rings are observed at 1623 cm$^{-1}$ ($\mathbf{8a}_{(py)}$) and 1598 cm$^{-1}$ ($\mathbf{8a}_{(Ph)}$), respectively. The band at 1338 cm$^{-1}$ corresponds to $\delta^s_{CH3}$. The out-of-plane bending vibrations $\gamma_{C=C}$, 11-$\gamma_{CH}$ of the benzene and pyridine rings are observed at 966 cm$^{-1}$, 836 cm$^{-1}$ and 765 cm$^{-1}$. Spectroscopic corroboration for the X-ray structure is obtained by application of the reducing difference procedure to polarized IR spectra, where the elimination of the bands at 1623 cm$^{-1}$, 1598 cm$^{-1}$, and 1338 cm$^{-1}$ in the same dichroic ratio is possible when the corresponding transition moments are colinearly oriented. On the other hand, the disappearance of the maxima at 966 cm$^{-1}$, 836 cm$^{-1}$, and 765 cm$^{-1}$ indicates the coplanar disposition of aromatic rings and double bonds and is in agreement with the crystallographic structure. As can be seen by the series of consequent eliminations of the IR bands of a given symmetry class obtained with reducing-difference IR-LD spectra, it is possible to obtain experimental data confirming the corresponding IR assignment and corroborating structural information.

IR characteristics of aromatic 1-methyl-4-[2-(4-hydroxyphenyl)ethenyl)]pyridinium] iodide and the corresponding 1-methyl-4-[2-(4-oxocyclohexadienyliden)ethyliden]-1,4-dihydropyridin will be strongly affected by the stabilization of the quinoide-like form in the second case. Conventional IR spectra are shown in Figure 6.10. In 1-methyl-4-[2-(4-hydroxyphenyl)ethenyl)]pyridinium] iodide the observed maxima at 3150 cm$^{-1}$ and 3021 cm$^{-1}$ can be assigned to stretching vibrations of the OH and CH groups in the 4′-position and the trans-disubstituted double bond, respectively. In the 1700 to 1400 cm$^{-1}$ spectroscopic range, there are significant differences compared with the IR spectrum of 1-methyl-4-[2-(4-oxocyclohexadienyliden)ethyliden]-1,4-dihydropyridine indicating the stabilization of the quinoide form after deprotonation of the OH group. The IR spectroscopic bands were assigned as follows: 1-methyl-4-[2-(4-hydroxyphenyl)ethenyl)]pyridinium] iodide, 1641 cm$^{-1}$ (8a, *py*), 1599 cm$^{-1}$ (8a, *ph*), 1560 cm$^{-1}$ ($\nu_{C=C}$); 1-methyl-4-[2-(4-oxocyclohexadienyliden)ethyliden]-1,4-dihydropyridine 1648 cm$^{-1}$ ($\nu_{C=C}$, *py*), 1616 cm$^{-1}$ ($\nu_{C=C}$, *ph*), and 1560 cm$^{-1}$ ($\nu_{C=C}$). The $\delta^s_{CH3}$ banding mode is at 1343 cm$^{-1}$ in 1-methyl-4-[2-(4-hydroxyphenyl)ethenyl)]pyridinium] iodide and at 1334 cm$^{-1}$ in the quinoide form. The positive charge in the N-atom is redistributed after the deprotonation of the OH group (Figure 6.3). The observed low-frequency shifting of the discussed maximum is expected and typical for other merocyanine dyes. For adequate structural elucidations of the compounds studied, it is important to perform the assignment of the out-of-plane modes. 1-Methyl-4-[2-(4-hydroxyphenyl)ethenyl)]pyridinium] iodide is characterized with bands at 989 cm$^{-1}$ ($\gamma_{=CH}$), 736 cm$^{-1}$ ($B_1$, py), and

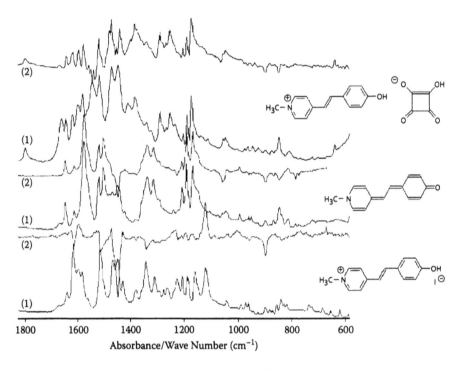

**FIGURE 6.10**   (1) Nonpolarized IR and (2) difference IR-LD spectra of compounds 1-methyl-4-[2-(4-hydroxyphenyl)ethenyl)]pyridinium] iodide, 1-methyl-4-[2-(4-hydroxyphenyl)ethenyl)] pyridinium] hydrogensquarate and 1-methyl-4-[2-(4-oxocyclohexadienyliden)ethyliden]-1,4-dihydropyridine.

833 cm$^{-1}$ (11-$\gamma_{CH}$, ph); and at 993 cm$^{-1}$, 972 cm$^{-1}$, and 774 cm$^{-1}$ in the corresponding IR spectrum of 1-methyl-4-[2-(4-oxocyclohexadienyliden)ethyliden]-1,4-dihydropyridine.

The elimination of the bands at 1641 cm$^{-1}$ in the IR-LD spectra of 1-methyl-4-[2-(4-hydroxyphenyl)ethenyl)]pyridinium] iodide leads to a disappearance of the bands at 1599 cm$^{-1}$ and 1560 cm$^{-1}$. This result indicates a planar structure of the molecule, because the presented elimination is possible when the discussed transition moments are mutually colinear. An additional confirmation of this conclusion follows from the observed elimination of the bands at 833 cm$^{-1}$, 736 cm$^{-1}$, and 989 cm$^{-1}$ at the same dichroic ratios. The simultaneous elimination of the bands of 1-methyl-4-[2-(4-oxocyclohexadienyliden)ethyliden]-1,4-dihydropyridine at 1648 cm$^{-1}$, 1618 cm$^{-1}$, 1560 cm$^{-1}$, and 1334 cm$^{-1}$ at equal dichroic ratios could also be realized in the planar geometry. The elimination of other bands at 993 cm$^{-1}$, 972 cm$^{-1}$, and 774 cm$^{-1}$ at the same dichroic ratio provides additional proof for the presented geometry (Figure 6.11).

The IR spectroscopic pattern of 1-methyl-4-[2-(4-hydroxyphenyl)ethenyl)]pyridinium] hydrogensquarate in the solid state is significantly complicated because of the presence of hydrogensquarate anions being characterized by a series of absorption peaks within the whole 1800 to 400 cm$^{-1}$ region. In this and other similar cases, Raman spectroscopy (Figure 6.12) supports the interpretation of the vibrational characteristics and structural analysis.

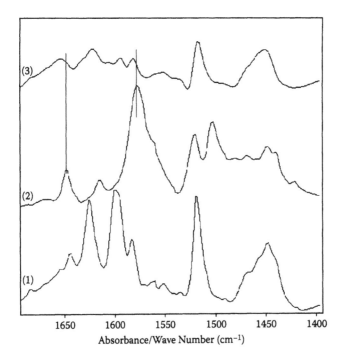

**FIGURE 6.11** Nonpolarized IR spectra of (1) 1-methyl-4-[2-(4-hydroxyphenyl)ethenyl)] pyridinium] iodide and (2) 1-methyl-4-[2-(4-oxocyclohexadienyliden)ethyliden]-1,4-dihydro-pyridin; (3) reduced IR-LD spectrum of 1-methyl-4-[2-(4-oxocyclohexadienyliden)ethyliden]-1,4-dihydropyridine after the elimination of the band at 1648 cm⁻¹.

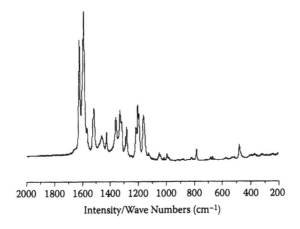

**FIGURE 6.12** Solid-state Raman spectrum of 1-methyl-4-[2-(4-hydroxyphenyl)ethenyl)] pyridinium] hydrogensquarate.

Surprisingly, the difference IR-LD spectra of 1-methyl-4-[2-(4-hydroxyphenyl)
ethenyl)]pyridinium] hydrogensquarate is characterized with a significant degree of
macroorientation of the suspended particles, which is typical for layered structures.
It is also typical for merocyanine dyes, but unusual for hydrogensquarates [215].
Absence of the $v_{OH}$ bands in the 3500 to 3200 cm$^{-1}$ region and the observation of
a broad band between 3000 and 2600 cm$^{-1}$ belonging to $v_{OH}$ of hydrogen bonded
hydrogensquarate ions indicates the participation of the OH function of a cation
in strong intermolecular interactions with anions. In the 1850 to 1500 cm$^{-1}$ region,
three new bands appear compared to the IR spectra: at 1805 cm$^{-1}$, 1662 cm$^{-1}$, and
1544 cm$^{-1}$ corresponding to the $v^s_{C=O}$, $v^{as}_{C=O}$, and $v_{C=C}$ stretches of hydrogensquarate.
The observed significant low-frequency shift for $v^{as}_{C=O(Sq)}$ of 38 cm$^{-1}$ from the typical hydrogensquarate values also supports the assumption of strong intermolecular
interactions of these species. The values of the aromatic fragments bands are practically uninfluenced in comparison to the typical values: 1644 cm$^{-1}$ (**8a**$_{(py)}$) and 1598
cm$^{-1}$ (**8a**$_{(Ph)}$). In the 850 to 750 cm$^{-1}$ range a high-frequency shifting of the B$_{1(py)}$ and
**11-**$\gamma_{CH}$ out-of plane modes with 72 cm$^{-1}$ and 13 cm$^{-1}$ is observed. This result indicates closely disposed and significantly $\pi$-interacted molecules in the layers, probably formed by dimers.

The elimination of the **8a**$_{(py)}$ band at 1644 cm$^{-1}$ leads to the disappearance (that
is, at equal dichroic ratios) of the band at 1598 cm$^{-1}$ (**8a**$_{(ph)}$) and $v^s_{C=O(Sq)}$ at 1805 cm$^{-1}$
and unambiguously defines the colinearity of corresponding transition moments in
the frame of the dimers. This geometry is also predicted theoretically to be the most
stable for the compound studied at the B3LYP/6-311++G** level of theory.

The planar geometry of the cations is also confirmed by the obtained reduction
of the 808 cm$^{-1}$ and 846 cm$^{-1}$ bands at relatively equal dichroic ratios. The 721 cm$^{-1}$
band also disappeared. Therefore, it belongs to the o.p. band of hydrogensquarate,
additionally supporting the assumptions for interactions of this type.

It is interesting to observe the typical multiple character of out-of-plane modes
of aromatic and hydrogensquarate fragments in solid-state IR spectra, which is
observed in the cases of significant $\pi$–$\pi$ interacted systems in the IR spectra of acetonitrile solutions. Moreover the discussed significant low-frequency shifting of the
of $v^{as}_{C=O(Sq)}$ with 30 cm$^{-1}$ and an absence of $v_{OH}$ bands in the 3500 to 3700 cm$^{-1}$ region
are typical for strong hydrogen bonding interactions. This means that the conclusion
for the aggregates is obtained by IR spectroscopy as well. It is interesting to note that
the IR spectrum of the sodium salt retains the discussed tendencies. The analysis
of corresponding 1-methyl-4-[2-(3-methoxy-4-hydroxyphenyl)ethenyl)]pyridinium]
hydrogensquarate monohydrate shows similar spectroscopic data and conclusions
about the aggregation processes in solution and in solid state. For the hydrogensquarate salts, the interpretation of the IR bands within the 1000 to 680 cm$^{-1}$ IR
region allow precise analysis; however, in the 1700 to 1450 cm$^{-1}$ region, the interpretation of the IR data is significant because of the strong overlapping effect.

Theoretically calculated IR spectra of the 1-methyl-4-[2-(4-hydroxyphenyl)ethe-
nyl)]pyridinium] cation and 1-methyl-4-[2-(4-oxocyclohexadienyliden)ethyliden]-1,4-
dihydropyridine support the elucidation of the experimental IR bands. The highest
frequency band in the IR spectrum of the cation at 3490 cm$^{-1}$ corresponds to the $v_{OH}$
mode (Figure 6.13). The low-intensive bands within the 3055 to 3000 cm$^{-1}$ range are

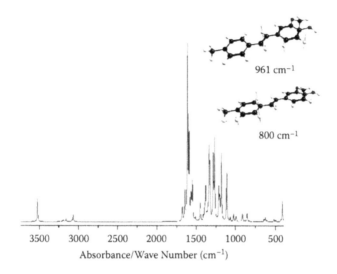

FIGURE 6.13   Theoretical IR-spectrum of 1-methyl-4-[2-(3-methoxy-4-hydroxyphenyl)
ethenyl)]pyridinium] cation and visualization of the selected $B_1$ transition moments.

assigned to CH stretching vibrations, while the maxima at 2965 cm$^{-1}$ and 2892 cm$^{-1}$
are assigned to CH$_3$ stretching vibrations. Within the whole 1650 to 1500 cm$^{-1}$ range
in-plane $A_1$ and $B_2$ modes of pyridine and benzene rings are observed: 1638 cm$^{-1}$
(**8a$_{py}$**), 1599 cm$^{-1}$ (**8a**, phenyl). The symmetric stretching C=C mode is at 1552 cm$^{-1}$;
$\gamma_{C=C}$ at 989 cm$^{-1}$; $B_1$ (**11-$\gamma_{CH}$**) at 734 cm$^{-1}$ (*Py*); and 833 cm$^{-1}$ (*Ph*). The theoretical IR
spectrum of the quinoide form is characterized with a disappearance of the typical
aromatic bands between the 1700 and 1500 cm$^{-1}$ range. The observed bands have
mixed character and are assigned as follows: 1657 cm$^{-1}$ ($\nu_{C=O}$), 1618 cm$^{-1}$ ($\nu_{C=Cpy}$),
and 1580 cm$^{-1}$ ($\nu_{C=C}$). The o.p. modes of the double bonds are predicted at 998 cm$^{-1}$
($\gamma_{=CH}$, cis-disubstituted), 972 cm$^{-1}$ ($\gamma_{=CH}$, trisubstituted). The bands for $\gamma_{C=O}$ and $\delta^s_{CH3}$
modes are at 774 cm$^{-1}$ and 1335 cm$^{-1}$, respectively. The last value is typical for $\delta^s_{CH3}$
in an N-CH$_3$ group.

The highest frequency band at 3523 cm$^{-1}$ corresponds to the $\nu_{OH}$ stretching vibration. In the solid state (see later) the corresponding IR maximum is shifted to a lower
frequency by more than 200 cm$^{-1}$ as a result of the participation of the OH group
in intermolecular hydrogen bonding. The bands within the 3000 to 3100 cm$^{-1}$ and
1640 to 1450 cm$^{-1}$ ranges belong to the i.p. $A_1$ and/or $B_2$ stretching vibrations of the
aromatic rings, and as a rule the pyridinium ones are at a higher frequency. The o.p.
$B_1$ bands are observed within the 900 to 680 cm$^{-1}$ range and the corresponding transition moments are perpendicularly oriented to the plane of the cation (Figure 6.13).

The nonpolarized IR spectrum shows a relatively intensive band at 3300 cm$^{-1}$
corresponding to the stretching $\nu_{OH}$ vibration of the hydrogen bonded OH group.
Bands at 1795 cm$^{-1}$, 1685 cm$^{-1}$, and 1589 cm$^{-1}$ correspond to $\nu^s_{C=O}$, $\nu^{as}_{C=O}$, and $\nu_{C=C}$
vibrations. Aromatic in-plane modes are at 1620 cm$^{-1}$ (**8a$_{(py)}$**) and 1600 cm$^{-1}$ (**8a$_{(Ph)}$**).
The theoretical values are 1618 cm$^{-1}$ and 1598 cm$^{-1}$, respectively. The band at 1338
cm$^{-1}$ corresponds to $\delta^s_{CH3}$. The absorption maxima belonging to the out-of-plane
bending vibrations $\gamma_{C=C}$, **11-$\gamma_{CH}$** benzene and of the pyridine rings are observed

at 960 cm$^{-1}$ (calc. 961 cm$^{-1}$), 854 cm$^{-1}$ (calc. 855 cm$^{-1}$), and 805 cm$^{-1}$ (calc. 800 cm$^{-1}$). Spectroscopic support for the X-ray structure is obtained by application of the reducing difference procedure to polarized IR spectra, where the elimination of the last bands at equal dichroic ratios is obtained, due to the corresponding transition moments being colinear. This result indicates the coplanar disposition of aromatic rings and double bonds and is in good agreement with the crystallographic structure.

1-Methyl-4-[2-(4-hydroxyphenyl)ethenyl)]pyridinium] hydrogenphosphate crystallizes in the monoclinic P2$_1$/c space group (Figure 6.14) and the molecules are joined into infinite parallel pseudolayers by strong intermolecular OH$\cdots$O-P hydrogen bonds (2.597 Å and OH$\cdots$O angle is 166.43°). Dihydrogenphosphate anions form infinite chains by strong (P)OH$\cdots$O(P) bonds with lengths of 2.630 Å (Figure 6.15). The obtained geometry parameters agree well with values of other stilbazolium salts [437–442]. Similar to other crystallographic studies of this class of organic compounds, crystallographic disorder of C7 and C8 in the dye leads to the observation of either a central double (C7–C8) or single bond (C7′–C8′) presumably due to a deviation of the highly π-conjugated system. Although deviation of the conjugation must presumably be intrinsic [437–442], the presence of a static disorder cannot be ruled out.

Compared with the crystallographic data of the corresponding hydrogensquarate, that is, 1-methyl-4-[2-(4-hydroxyphenyl)ethenyl)]pyridinium] hydrogensquarate [215] (Figure 6.16); it was found that both of the structures crystallized in the same monoclinic space group P2$_1$/c. The cations and anions also form layer structures by hydrogen bonding of the type (Figure 6.17) OH$\cdots$O$_{(Sq)}$ (2.690 Å) and $_{(Sq)}$OH$\cdots$O$_{(Sq)}$ (2.545 Å), respectively.

1-Methyl-4-[2-(3-methoxy-4-hydroxyphenyl)ethenyl)]pyridinium] hydrogensquarate monohydrate crystallizes in the centrosymmetric space group P-1

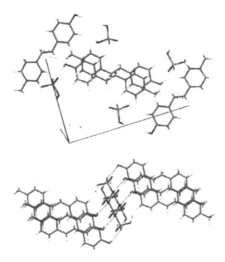

**FIGURE 6.14**   Unit cell and hydrogen bonding of 1-methyl-4-[2-(4-hydroxyphenyl)ethenyl)] pyridinium] hydrogenphosphate.

(Figure 6.18). Similar to 1-methyl-4-[2-(4-hydroxyphenyl)ethenyl)]pyridinium] hydrogenphosphate, infinite layers are formed by intermolecular hydrogen bonds of the types OH···O=C$_{(Sq)}$ (2.684 Å) and $_{(Sq)}$OH···OH$_2$ (2.575 and 2.850 Å) (Figure 6.19). The interlayer interaction is the basis of the obtained optical properties of this compound. One of the hydrogensquarate anions is disordered, which is typical for systems containing this counterion (Figure 6.18). The geometry of the cation is effectively flat with a maximal deviation of 0.5(8)°. The OCH$_3$ group lies in the plane of 1-methyl-4-[2-(4-hydroxyphenyl)ethenyl)]pyridinium] fragment with an angle (H$_3$)COCC(H)

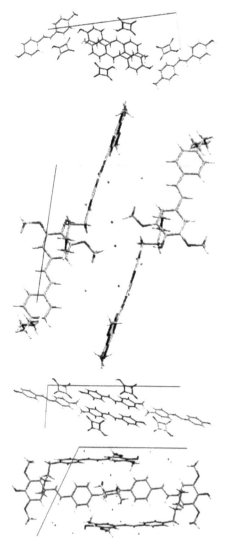

**FIGURE 6.15**   View along the $a$, $b$, and $c$ axes of (1) 1-methyl-4-[2-(4-hydroxyphenyl)ethe-nyl)]pyridinium] hydrogenphosphate and (2) 1-butyl-4-[2-(3,5-dimethoxy4-hydroxyphenyl) ethenyl)]piridinium] chloride tetrahydrate. **(See color insert.)**

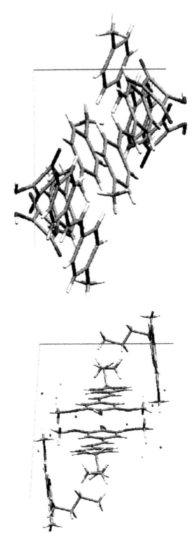

**FIGURE 6.15**   (Continued) View along the *a*, *b*, and *c* axes of (1) 1-methyl-4-[2-(4-hydroxy-phenyl)ethenyl)]pyridinium] hydrogenphosphate and (2) 1-butyl-4-[2-(3,5-dimethoxy4-hy-droxyphenyl)ethenyl)]piridinium] chloride tetrahydrate. (**See color insert.**)

of 4.7(1)°; the unit cell contains two coplanar cations at a distance of 3.460 Å, thus causing a colinear orientation of the out-of-plane $B_1$ transition moments of the aromatic fragments. Remarkably this compound possesses NLO properties for crystals of nanoparticle size (Figure 6.20).

Comparing the crystallographic data for the bond lengths of the aromatic and quinoide-like forms of the 1-methyl-4-[2-(4-hydroxyphenyl)ethenyl)]pyridinium] salts (Table 6.5), it can be seen that the bond lengths differ within 0.032 to 0.004 Å. This means that the data for the bond lengths cannot be used for an unambiguous assignment of the aromatic or quinoide form in these compounds. Moreover, in the crystal

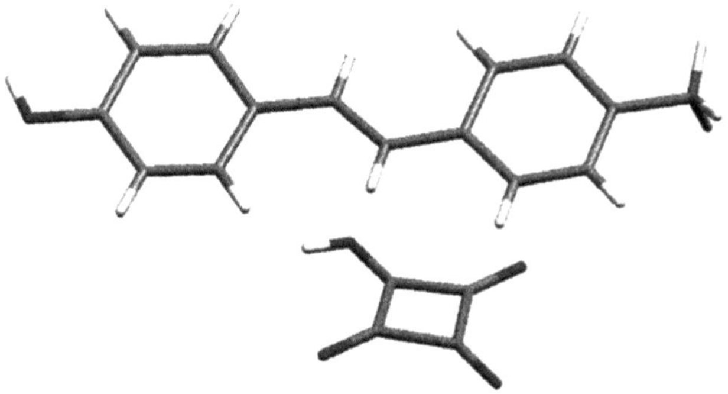

**FIGURE 6.16** Crystal structure of 1-methyl-4-[2-(4-hydroxyphenyl)ethenyl)]pyridinium] hydrogensquarate.

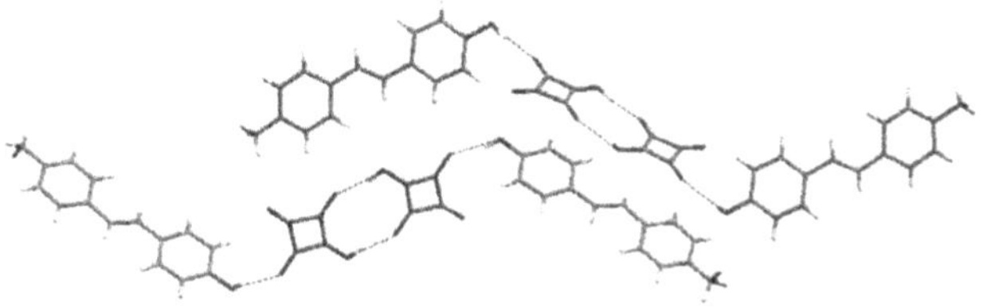

**FIGURE 6.17** Hydrogen bonding in 1-methyl-4-[2-(4-hydroxyphenyl)ethenyl)]pyridinium] hydrogensquarate.

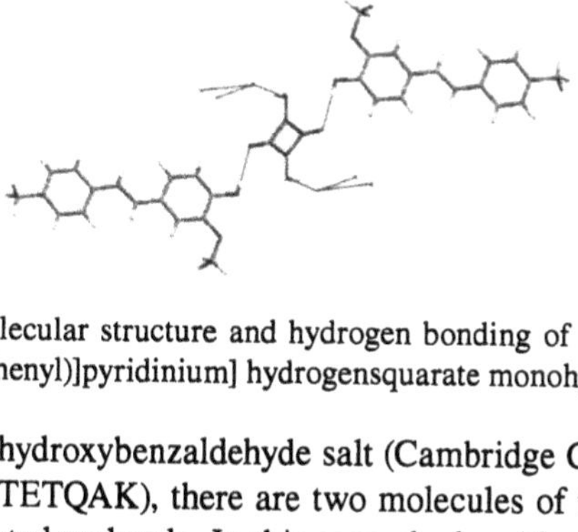

**FIGURE 6.18** The molecular structure and hydrogen bonding of 1-methyl-4-[2-(3-methoxy-4-hydroxyphenyl)ethenyl)]pyridinium] hydrogensquarate monohydrate.

structure of the 2,4-dihydroxybenzaldehyde salt (Cambridge Crystallographic Data Centre [CCDC] code TETQAK), there are two molecules of the dye, a protonated cation and a deprotonated molecule. In this case, the bond lengths are significantly changed by 0.007 to 0.019 Å. In such cases, the polarized IR spectroscopy gives direct and precise data for the single-charge redistribution in the molecules.

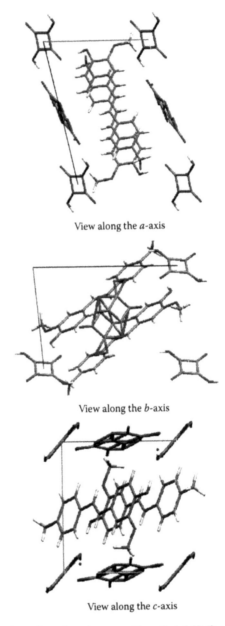

View along the *a*-axis

View along the *b*-axis

View along the *c*-axis

**FIGURE 6.19**    View along the *a, b,* and *c* axes of 1-methyl-4-[2-(3-methoxy-4-hydroxyphe-nyl)ethenyl)]pyridinium] hydrogensquarate monohydrate.

In the cases with the long chain N-aliphatic substituents, for example, 1-butyl-4-[2-(3,5-dimethoxy-4-hydroxyphenyl)ethenyl)]pyridinium] chloride tetrahydrate (Figure 6.21), crystallizing in triclinic P-1 space group, the unit cell contains two independent 1-butyl-4-[2-(3,5-dimethoxy4-hydroxyphenyl)ethenyl)]pyridinium] cat-ions that differ from the butyl chain ($H_3CCH_2CH_2CH_2$) with torsion angles of 80.0(9)°

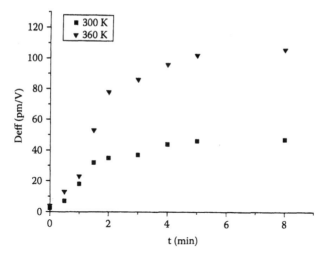

**FIGURE 6.20** Dependence of the second-order susceptibilities of 1-methyl-4-[2-(3-methoxy-4-hydroxyphenyl)ethenyl)]pyridinium] hydrogensquarate monohydrate at 1064 nm versus time with frequency repetition of about 1 kHz. (**See color insert.**)

**TABLE 6.5**
**Bond Lengths [Å] of the Quinoide-Like and Aromatic Forms of 1-methyl-4-[2-(4-hydroxyphenyl)ethenyl)]pyridinium] Salts, Using the Atom Numbering Scheme in Figure 6.1**

| | | CCDC Codes of the Structures[a] | | | |
|---|---|---|---|---|---|
| Name | | TETQAK | | | |
| Definition | FETPUQ | Molecule 1 | Molecule 2 | HEZZER | JOZBEE |
| $N_1$-$C_2$ | 1.333 | 1.343 | 1.345 | 1.344 | 1.337 |
| $C_2$-$C_3$ | 1.325 | 1.357 | 1.358 | 1.365 | 1.365 |
| $C_3$-$C_4$ | 1.366 | 1.394 | 1.407 | 1.392 | 1.400 |
| $C_4$-$C_5$ | 1.393 | 1.396 | 1.401 | 1.396 | 1.400 |
| $C_5$-$C_6$ | 1.383 | 1.366 | 1.355 | 1.358 | 1.367 |
| $C_6$-$N_1$ | 1.349 | 1.349 | 1.356 | 1.348 | 1.346 |
| $C_4$-$C_7$ | 1.472 | 1.449 | 1.445 | 1.450 | 1.451 |
| $C_7$-$C_8$ | 1.298 | 1.338 | 1.336 | 1.336 | 1.339 |
| $C_8$-$C_{1'}$ | 1.428 | 1.450 | 1.443 | 1.451 | 1.457 |
| $C_{1'}$-$C_{2'}$ | 1.419 | 1.397 | 1.407 | 1.403 | 1.408 |
| $C_{2'}$-$C_{3'}$ | 1.361 | 1.373 | 1.369 | 1.380 | 1.380 |
| $C_{3'}$-$C_{4'}$ | 1.463 | 1.380 | 1.400 | 1.411 | 1.402 |
| $C_{4'}$-O(N) | 1.253 | 1.349 | 1.330 | 1.373 | 1.378 |
| $C_{4'}$-$C_{5'}$ | 1.461 | 1.396 | 1.400 | 1.400 | 1.351 |
| $C_{5'}$-$C_{6'}$ | 1.345 | 1.366 | 1.384 | 1.379 | 1.383 |
| $C_{6'}$-$C_{1'}$ | 1.405 | 1.403 | 1.402 | 1.395 | 1.392 |

[a] CCDC—Cambridge Crystallographic Database Centre, http://www.ccdc.cam.ac.uk/.

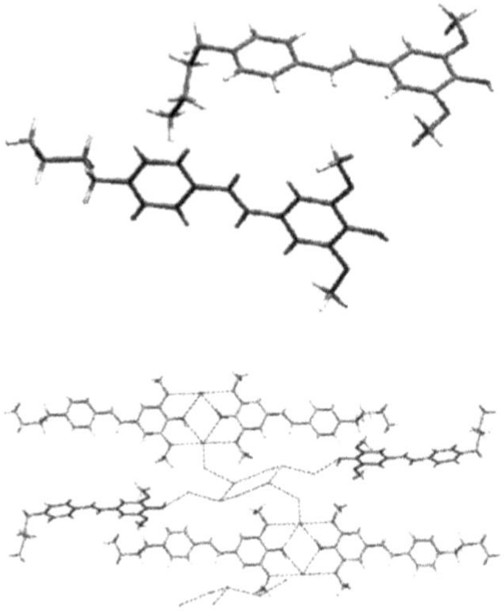

**FIGURE 6.21** The molecular structure and linkage of 1-butyl-4-[2-(3,5-dimethoxy4-hydroxyphenyl)ethenyl)]pyridinium] chloride tetrahydrate.

and 173.6(3)°, respectively (Figure 6.21). The molecules are joined into infinite layers, formed by two different dimers of the cations and anions, and include solvent molecules. The observed hydrogen bonds are OH···$OH_2$ (2.814 Å), HOH···$O(CH_3)$ (2.960 Å), OH···Cl (2.967 Å), HOH···$Cl^-$ (3.034, 3.188, 3.161, and 3.062 Å) and HOH···$OH_2$ (2.772 Å) (Figure 6.21). The 4-[2-(3,5-dimethoxy-4-hydroxypheny l)ethenyl)]piridinium] fragment of the cation is flat with a deviation from a total planarity of 0.2°. Surprisingly, although the typical crystallographic disorder of the double bond in the dye leads to the observation of the isolated single and double bonds, that is, to a deviation of the highly $\pi$-conjugated system, only disorder of the anions is observed in this case. Like other dyes [215], the disorder leads to a relatively high R factor.

For the precise analysis of the conformational preference in this type of system, the theoretical conformational analysis gives a reasonable explanation for the gas phase. For the cationic and anionic structure of the latter dye, the data show that in both cases conformers are stabilized in which the geometry of the butyl chain with torsion angles of 178.9(1)° and 179.9(0)° is preferred (Figure 6.27). These data suggest that the observation of the conformer in the solid state with a torsion angle of the discussed chain of 80.0(9)° is a result of the geometry effects in the condensed phase. In these systems the theoretical vibrational analysis is performed using the geometry of the compounds obtained by single-crystal X-ray diffraction (if the data are available) as an input file.

The nonpolarized IR spectrum shows a relatively intensive band at 3383 cm$^{-1}$ corresponding to the stretching $v_{OH}$ vibration of the hydrogen bonded OH group in merocyanine dyes, thus correlating to the obtained crystallographic data. The band at 3515 cm$^{-1}$ belongs to the same vibration but is due to the included solvent molecules. In the 1700 to 1600 cm$^{-1}$ range, the aromatic i.p. modes of the benzene and pyridine rings are at 1614 cm$^{-1}$ (**8a$_{(py)}$**) and 1584 cm$^{-1}$ (**8a$_{(Ph)}$**). The absorption maxima belonging to the o.p. bending vibrations $\gamma_{C=C}$, $11-\gamma_{CH}$ of benzene and the o.p. mode of the pyridine rings are observed at 996 cm⁻¹, 849 cm⁻¹, and 734 cm⁻¹. The spectroscopic corroboration of the crystallographic structure is obtained by application of the reducing difference procedure to the polarized IR spectra, where the elimination of the bands at 1614 cm⁻¹ and 1584 cm⁻¹ at the same dichroic ratios is possible when the corresponding transition moments are colinear. On the other hand, the disappearance of the maxima at 996 cm⁻¹, 849 cm⁻¹, and 734 cm⁻¹ at the same dichroic ratios (Figure 6.22) indicate a mutually coplanar disposition of the aromatic rings and the double bond, also in good agreement with the crystallographic structure. This result is similar to those for other derivatives with small $R_1$ substituents (Figure 6.1).

In the corresponding Raman spectrum, the series of i.p. vibrations of the benzene and pyridine rings are at 1616, 1584, 1561, and 1516 cm⁻¹. The o.p. modes at 996 and 739 cm⁻¹ are characterized by relatively low intensity, while the band at 849 cm⁻¹ is absent.

Polarized IR spectroscopy can also be successfully applied to quinoline dyes, such as the derivatives depicted in Figure 6.23. In these cases, crystallographic data are for the three known structures shown in Table 6.6. In these cases, the polarized IR spectroscopy in the solid state appeared to be especially attractive.

In the case of n-butyl-N-4-(N-dimethylaminostyryl)quinolinium iodide we have a chance to compare the IR-LD spectroscopic data with the crystallographic data, thus confirming the validity of the conclusions about the structure of the compound obtained by the IR spectroscopy. n-Butyl-N-4-(N-dimethylaminostyryl)quinolinium

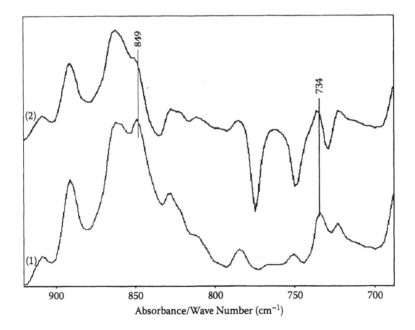

**FIGURE 6.22** (1) Nonpolarized IR and (2) reduced IR-LD spectrum of 1-butyl-4-[2-(3,5-dimethoxy4-hydroxyphenyl)ethenyl)]piridinium] chloride tetrahydrate after the elimination of the band at 849 cm⁻¹.

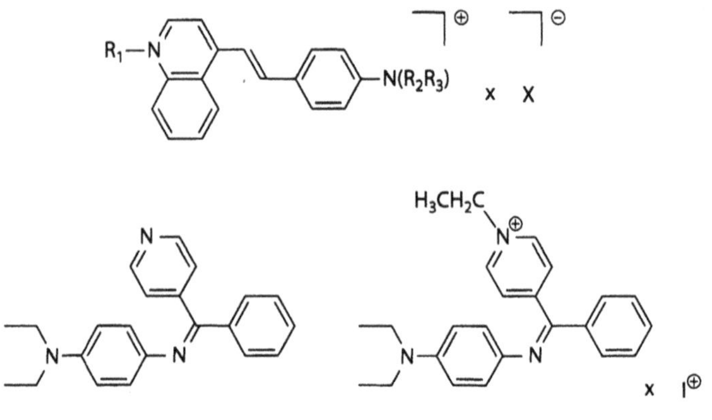

**FIGURE 6.23**  Chemical diagram of the salts.

---

**TABLE 6.6**
**Crystal Structures of the Salts with Formulae, Given in Figure 6.23**

| $R_1$ | $R_2$ | $R_3$ | X | CCDC Code[a] |
|-------|-------|-------|---|-----------|
| $CH_3$ | $CH_3$ | $CH_2CH_2OH$ | I⁻ | ABILIG |
| $CH_3$ | $CH_3$ | $CH_3$ | Toluene-p-sulfonate | FEDRIQ |
| $CH_3$ | $CH_3$ | $CH_3$ | hexamolybdate | ZAGPUS |

[a] CCDC—Cambridge Crystallographic Database Centre, Cambridge Structural Database (CSD) version 5.29 updates (January 2008), http://www.ccdc.cam.ac.uk/.

---

iodide crystallizes in the triclinic P-1 space group and the unit cell contains two n-butyl-N-4-(n-dimethylaminostyryl)quinolinium cations and two iodide anions. The cation is effectively flat with a deviation of total planarity of 5.7°.

The performed conformational analysis of the cationic form of the compound shows that its energetically favorable conformer exhibits with the geometry of the butyl chain with a torsion angle of 179.3(2)°. The data suggest that the observation of a conformer in the solid state with a torsion angle of the discussed chain of 90.4(7)° is a result of the packing effects in the condensed phase. The obtained four IR bands within the IR region of 1750 to 1550 cm⁻¹ correspond to the in-plane vibrations of both the aromatic fragments and correlate well with the theoretical data. Independent of this, the unit cell of the compound studied contains two cations and their coplanar orientation leads to the elimination of these bands at equal dichroic ratios.

Anyles are another class of dyes successfully studied by polarization IR spectroscopy of oriented colloids. N,N-Diethyl-N'-[(Z)-phenyl(pyridin-4-yl)methylidene]benzene-1,4-diamine and its N-ethyl pyridinium iodide (Figure 6.23) will be presented next.

N,N-diethyl-N'-[(Z)-phenyl(pyridin-4-yl)methylidene]benzene-1,4-diamine crystallizes in the P-1 space group and its unit cell contains two pairs of nonequivalent molecules, differing in the dihedral angle values of 100.9(0)° and 96.0(7)° for the $(CH_3)CH_2NCH_2CH_3$ angle. Only a short contact between two neighboring pyridinium fragments is found with the length of 3.16(7)°. The planes of the aromatic fragment are mutually oriented at angles of 57.8(5)°, 71.6(9)°, and 66.0(4)°, thus preventing any type of conjugation. According to the Bethe theory, the different orientation of these molecules will affect the elimination of the subcomponents of the given IR bands at different dichroic ratios (Figure 6.24).

However, the effect of the N-ethylation is easily studied by IR spectroscopy. Comparing its IR spectra in the solid state to that of the 4-amino derivative within the region of 1700 to 1450 cm$^{-1}$, it is apparent that the N-ethylation does not significantly affect the i.p. characteristics of the benzene rings and pyridinium fragment (bands about 1610 cm$^{-1}$, 1592 cm$^{-1}$, 1581 cm$^{-1}$, 1540 cm$^{-1}$, and 1510 cm$^{-1}$). As a rule, the second structural fragment is characterized with high-frequency values of the discussed vibrations. Only a difference of less than 2 cm$^{-1}$ is obtained by comparing the data of N,N-diethyl-N′-[(Z)-phenyl(pyridin-4-yl)methylidene]benzene-1,4-diamine and its N-ethyl pyridinium iodide. In the IR spectrum of the salt a well-defined band at 1641 cm$^{-1}$ corresponding to the $v_{C≡N}$ stretching vibration (theoretical value is 1640 cm$^{-1}$) is observed. In the former compound, the discussed maximum is low frequency shifted by 7 cm$^{-1}$. The application of the reducing-difference procedure for polarized IR-LD spectra shows that on elimination of the i.p. bands of the pyridinium fragment maxima with same symmetry class of the other two aromatic fragments are observed in both cases as a result of their mutual cross-orientation in each of the molecules in N,N-diethyl-N′-[(Z)-phenyl(pyridin-4-yl)methylidene]benzene-1,4-diamine and its N-ethyl pyridinium iodide.

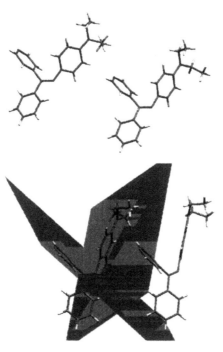

**FIGURE 6.24** Linkage and mutual orientation of the aromatic fragment in the molecule of N,N-diethyl-N′-[(Z)-phenyl(pyridin-4-yl)methylidene]benzene-1,4-diamine.

The IR spectroscopic region within 900 and 600 cm$^{-1}$ is characterized with a series of o.p. IR bands of the three aromatic fragments: at 860 cm$^{-1}$ (o.p. vibration of pyridine fragment), 820 cm$^{-1}$ (p-disubstituted benzene ring), and 721/705 cm$^{-1}$ (monosubstituted benzene ring). The effect of N-ethylation leads to a low frequency shifting of 9 cm$^{-1}$ for the o.p. band of the pyridinium fragment. Similar to the i.p., the o.p. bands are eliminated at different dichroic ratios as a result of the cross-orientation of the benzene rings and pyridine fragments in both of the compounds.

## 6.2   DICYANOISOPHORONE DERIVATIVES

Organic NLO materials have been intensively studied recently due to their potential application in various fields, such as telecommunications and optical information processes. A series of dicyanoisophorones have been synthesized and structurally determined by single-crystal X-ray diffraction by Kolev and coauthors [421–428] on the basis of Lemke's [444] pioneering studies. Experimental and computational studies on the vibrational spectra of 2-[5,5-dimethyl-3-(2-phenyl-vinil)-cyclohex-2-enylidene]-malononitrile have been reported [445]. However, the presence of aromatic and conjugated systems in the structures of these compounds makes the assignment of individual bands difficult using IR spectra. Therefore, this chapter deals with the solid-state IR-LD analysis of 2-[5,5-dimethyl-3-(2-phenyl-vinyl)-cyclohex-2-enylidene]-malononitrile (*1*), 2-{5,5-dimethyl-3-[2-(2-methoxyphenyl)vinyl]cyclo-hex-2-enylidene}malononitrile (*2*), and 2-[3-[2-(2,4-dimethoxyphenyl)

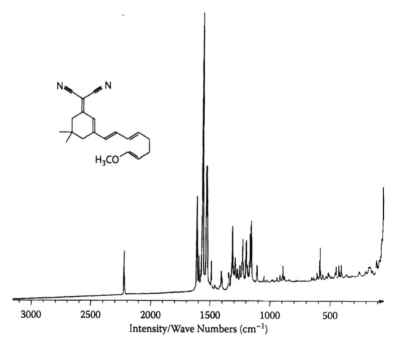

**FIGURE 6.25**   Solid state Raman spectrum of 2-[3-[2-(2-methoxyphenyl)vinyl]-5,5-dimethylcyclohex-2-enylidene]malononitrile.

vinyl]-5,5-dimethylcyclohex-2-enylidene]malononitrile (*3*) (Figure 6.26 through Figure 6.28).

All these compounds contain a common conjugated structural fragment, thus leading to similar IR spectral characteristics as only the substituents in the aromatic fragment are different. For this reason, the nonpolarized IR spectra of the three compounds are similar in the 1700 to 1350 cm$^{-1}$ region, indicating $\nu_{C=C}$ stretching frequencies and radial vibrations of aromatic rings. IR characteristic band assignments were carried out by comparison with the corresponding Raman bands (Figure 6.25), and correlate well with the theoretical ones. All the compounds are characterized with $\nu_{C\equiv N}$ stretching peaks at 2200 cm$^{-1}$, but in some the cases a multiple character is observed, making assignment difficult by conventional IR spectroscopy. However, in the 1000 to 680 cm$^{-1}$ frequency region, IR characteristics for out-of-plane aromatic modes as well as of the unsaturated conjugated system are different as a result of different types of substituted aromatic fragments. On the other hand, the strong overlapping effects in the latter region make a detailed IR band assignment with conventional IR spectroscopy difficult as well.

Difference IR-LD spectra of all three compounds indicate a significant degree of macroorientation of guest molecules in liquid crystal medium. In the case of 2-[3-[2-(2-methoxyphenyl)vinyl]-5,5-dimethylcyclohex-2-enylidene]malononitrile the difference IR-LD spectrum shows a simultaneous elimination of both $\nu_{C\equiv N}$ components at 2200 cm$^{-1}$, thus indicating a crystal field splitting effect of the $\nu_{C\equiv N}$ maximum. The application of the reducing difference procedure for polarized spectra interpretation leads to the following results. For 2-[3-[2-(phenyl)vinyl]-5,5-dimethylcyclohex-2-enylidene]malononitrile (Figure 6.26), the elimination of the 1616 cm$^{-1}$ peak leads to disappearance of the 1596 cm$^{-1}$ (8a, i.p. phenyl mode, A$_1$), 1558 cm$^{-1}$, 1525 cm$^{-1}$ and 1496 cm$^{-1}$ peaks (19a, phenyl, i.p.). For 2-[3-[2-(2-methoxyphenyl)vinyl]-5,5-dimethylcyclohex-2-enylidene]malononitrile there was simultaneous vanishing of the **8a** and **19a** peaks at 1592 cm$^{-1}$ and 1486 cm$^{-1}$ and a strong reduction of the $\nu_{C=C}$ ones at 1616 cm$^{-1}$, 1560 cm$^{-1}$, and 1527 cm$^{-1}$ (Figure 6.27), which can be explained by a deviation from colinearity of the A$_1$ and $\nu_{C=C}$ transition moments. The reduction of the 1608 cm$^{-1}$ peak in the IR-LD spectra of 2-[3-[2-(2,4-dimethoxyphenyl)vinyl]-5,5-dimethylcyclohex-2-enylidene]malononitrile (*3*) leads to an intensity decreasing of maxima at 1596 cm$^{-1}$, 1569 cm$^{-1}$, 1527 cm$^{-1}$, and 1502 cm$^{-1}$ (Figure 6.28). This data indicates flat molecules and colinear orientation of the three C=C bonds, a fact established by X-ray diffraction of the corresponding methoxy derivatives. On the other hand, in the frame of the unit cell of (*3*), the mutual orientation of the neighboring molecules caused the colinearity of A$_1$ phenyl modes with $\nu_{C=C}$ stretch ones thus determining the simultaneous disappearance of all discussed maxima in the reduced IR spectra. In 2-[3-[2-(2-methoxyphenyl)vinyl]-5,5-dimethylcyclohex-2-enylidene]malononitrile (*2*), both molecules in the unit cell are inclined at an angle of 65.0(6)°, which indicates that the elimination of out-of-plane modes (B$_1$) will lead to the observation of secondary peaks assigned to other molecules (see later). Conversely, the unit cell of 2-[3-[2-(2,4-dimethoxyphenyl)vinyl]-5,5-dimethylcyclohex-2-enylidene]malononitrile (*3*) contains parallel molecules, which cause the total disappearance of all o.p. maxima. The elimination of the monosubstituted phenyl ring out-of-plane mode (B$_1$) at 751 cm$^{-1}$ (11-$\gamma_{CH}$) resulted in the vanishing of peaks of 4-$\gamma_{Ar}$ at 962 cm⁻¹ and 721 cm⁻¹, all of which are negatively oriented in the difference IR-LD spectrum. As far as the latter peaks could be attributed to o.p. $\gamma_{=CH}$ modes of trans di- and trisubstituted C=C bonds, these data are in accordance with a flat configuration of conjugated and aromatic systems in the molecule of (*I*). Similar results were established as well as for the other compounds, where in (*2*), for instance, the simultaneous disappearance of o-disubstituted benzenes peak at 764 cm⁻¹ and 970 cm⁻¹ was determined. The procedure led to second peaks at 760 cm⁻¹ and 963 cm⁻¹, corresponding to a second molecule included in the unit cell of 2-[3-[2-(2-methoxyphenyl)vinyl]-5,5-dimethylcyclohex-2-enylidene]malononitrile     and oriented at an angle of 65.0(6)°. In the case of (*3*), the o.p. mode for *m*-trisubstituted phenyls at 829 cm⁻¹ disappeared at the same dichroic ratio as the 966 cm⁻¹ peak, thus determining the assignment of the maximum as $\gamma_{=CH}$ of the trans-disubstituted C=C bond. The total reduction of both o.p. bands correlated with single-crystal X-ray data of 2-[3-[2-(2,4-dimethoxyphenyl)vinyl]-5,5-dimethylcyclohex-2-enylidene]malono-nitrile, where the planar configuration of both molecules in the unit cell has been established (Figure 6.26 through Figure 6.28).

Remarkable NLO properties have been obtained from the 3-dicyanomethylen-5,5-dimethyl-1-[2-(4-hydroxyphenyl)ethenyl)]-cyclohexene derivatives [419].

The presence of possible resonance forms depending on the solvent polarity and pH explained the recorded UV-Vis data. Solvatochromic effects within 45 to 70 nm are obtained for the forms in Figure 6.29, depending on the solvent polarity ($H_2O$ to acetone or $CHCl_3$) (Figure 6.30). The effect could be explained with changes in the electron density distribution. However, the anionic form, stabilized as the Na-salt,

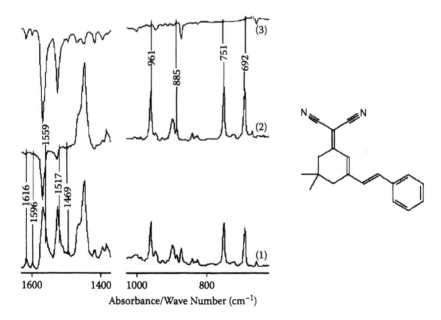

**FIGURE 6.26** (1) Nonpolarized IR and reduced IR-LD spectra of 2-[3-[2-(phenyl)vinyl]-5,5-dimethylcyclohex-2-enylidene]malononitrile (*I*) after the elimination of the peaks at (2) 1616 cm⁻¹ and (3) 751 cm⁻¹.

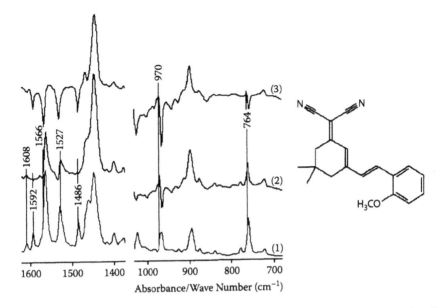

**FIGURE 6.27**   (1) Nonpolarized IR and reduced IR-LD spectra of 2-[3-[2-(2-methoxyphenyl) vinyl]-5,5-dimethylcyclohex-2-enylidene]malononitrile (2) after the elimination of the peaks at (2) 1592 cm⁻¹ and (3) 764 cm⁻¹.

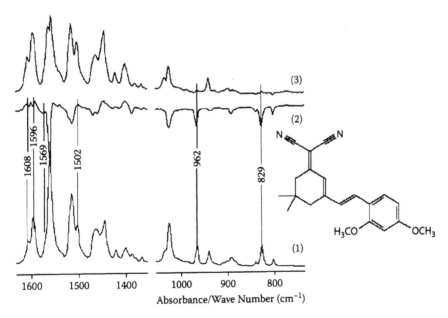

**FIGURE 6.28**   (1) Nonpolarized IR and reduced IR-LD spectra of 2-[3-[2-(2,4-dimethoxy-phenyl)vinyl]-5,5-dimethylcyclohex-2-enylidene]malononitrile (3) after the elimination of the peaks at (2) 1608 cm⁻¹ and (3) 829 cm⁻¹.

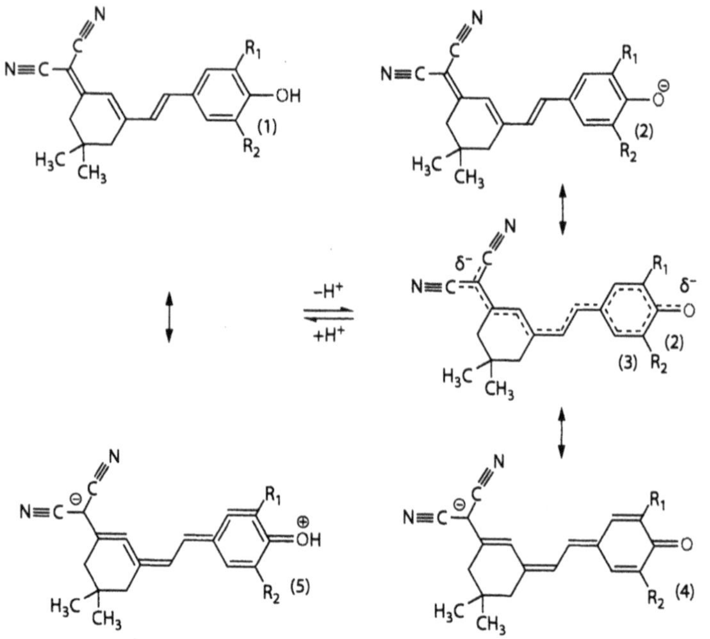

**FIGURE 6.29** Chemical diagram of different resonance forms compounds with $R_1 = R_2 = H$, $R_1 = H$, $R_2 = OCH_3$, and $R_1 = R_2 = OCH_3$. **(See color insert.)**

is characterized by significant solvatochromic effects of up to 135 nm (CHCl$_3$ (473 nm) and C$_2$H$_5$OH (572 nm), which are pronounced in polar protic solvents (C$_2$H$_5$OH and CH$_3$OH). In a CH$_3$CN solution, the second band (633 nm) could be explained with aggregation. The data showed that polar protic solvents stabilized the quinoide form of the anionic 4-hydroxy derivative (Figure 6.29), because the $\lambda_{max}$ shifted from 442 nm to 572 nm after deprotonation. Direct evidence for the latter assumption is obtained by a comparison of the IR spectra of the aromatic and quinoide-like forms (Figure 6.31); the aromatic characteristic IR bands of the neutral form disappeared and a low-frequency shifting of the $v_{C=C}$ maxima with 123 cm$^{-1}$ was observed. Like UV spectroscopy, theoretically calculated NLO properties confirm pronounced NLO properties of the anionic form in comparison to the neutral one (Figure 6.32). The $\beta$ value is 2.5 times higher than the neutral form and 6.2 times higher than p-nitroaniline, a charismatic NLO material.

The NLO properties in nanoparticle sizes have been obtained [419] (Figure 6.32). From the reported data one can see that decreasing nanocrystalline sizes favor enhanced second-order susceptibilities. For all the samples the optimal ratio between the fundamental and writing beams was equal to about 15. For the samples larger than 300 nm, the effect was independent of the nanoparticle sizes. The relaxation of the optically formed nonlinear grating was not less than 80%. The obtained effective optical susceptibilities are comparable to those of the excellent borate crystals.

The IR spectral data for the analogous methoxy derivatives are presented in Figure 6.33.

Difference IR-LD spectrum of quinoide forms (*1*) in the solid state indicates a significant degree of macroorientation of guest molecules in liquid crystal medium and allow adequate application of the reducing difference procedure. According to the general principles, the colinearity of given transition moments result in the elimination of corresponding bands at equal dichroic ratios. Simultaneous elimination of bands at 1604 cm$^{-1}$, 1554 cm$^{-1}$, 1517 cm$^{-1}$, 1590 cm$^{-1}$, and 1508 cm$^{-1}$ in the 3,5-dimethoxy-4-hydroxy derivative is in accordance with crystallographic data showing the colinearity of the $\nu_{C=C}$ stretching modes of trans disubstituted, trisubstituted, and tetrasubstituted double bonds in the frame of one molecule. In-plane aromatic modes p-disubstituted phenyl (A$_1$), that is, the **8a** and **19a** modes, are also

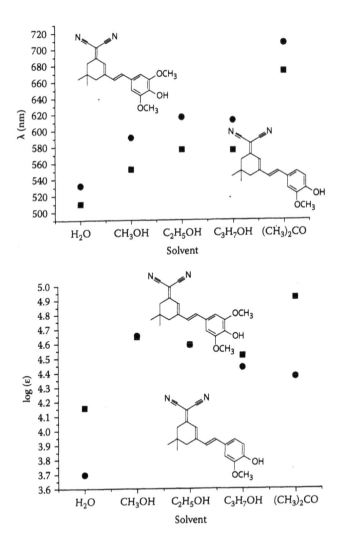

**FIGURE 6.30** Electronic spectra of different substituted dicyanoisophorones. (**See color insert.**)

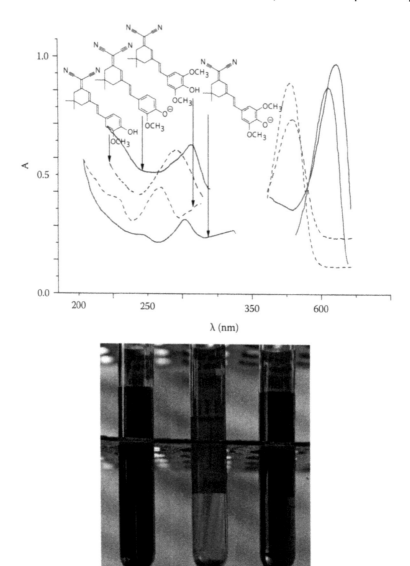

**FIGURE 6.30** (Continued) Electronic spectra of different substituted dicyanoisophorones. **(See color insert.)**

colinear in the isolated molecule, but in the unit cell they are colinear to the afore-mentioned transition moments. In this reduction, however, second pairs of peaks occur belonging to the same modes of other differently oriented molecules in the unit cell. The obtained intermolecular OH···NC bond resulted in the observation of a low-frequency maximum $v_{OH}$. The elimination of the bands at 958 cm$^{-1}$, 721 cm$^{-1}$, 841 cm$^{-1}$, and a broad one at 647 cm$^{-1}$ indicate their $\gamma_{CH}$ trisubstituted and trans-disubstituted, 11-$\gamma_{CH}$ (p-disubstituted phenyl) and $\gamma_{COH}$ o.p. character, due to the planar geometry of the molecule obtained both experimentally and theoretically

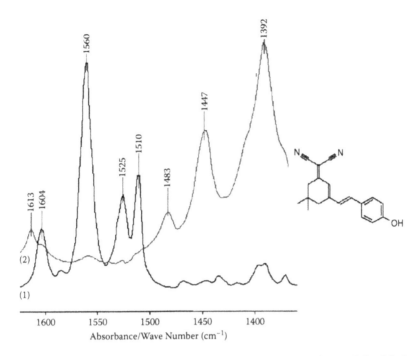

**FIGURE 6.31**   IR spectrum of (1) aromatic and (2) quinoide-like forms of the 4-hydroxy derivative.

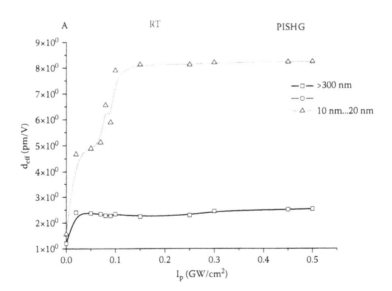

**FIGURE 6.32**   Dependence of the second-order optical susceptibility for the sample 2-[3-[2-(4-hydroxyvinyl]-5,5-dimethylcyclohex-2-enylidene]malononitrile versus the effective photoinducing pump power during simultaneous treatment by two coherent laser beams with a ratio of the fundamental to the writing beam ratio equal to about 15. (**See color insert.**)

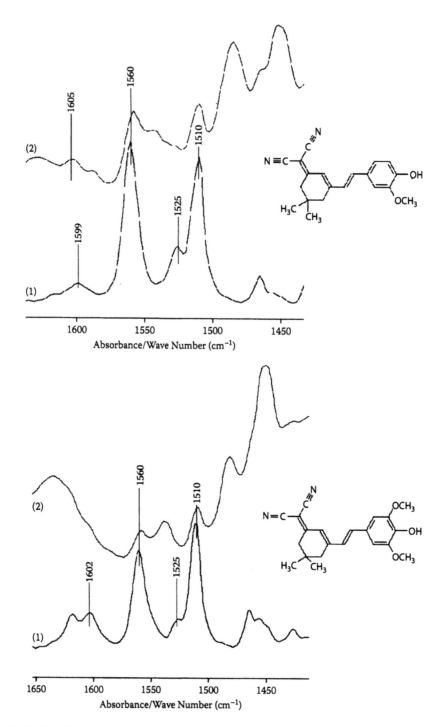

**FIGURE 6.33** IR spectra of the 3-methoxy-4-hydroxy and 3,5-dimethoxy-4-hydroxy derivates and (1) and (2) their Na-salts dissolved in solvent mixture $C_4H_8O:C_2H_5OH$ 8:2.

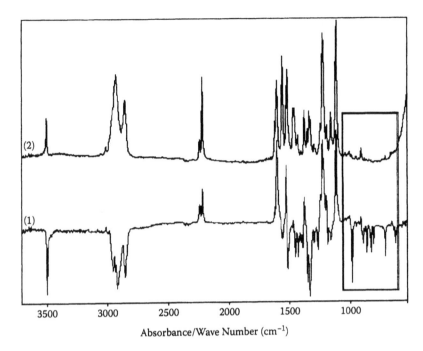

**FIGURE 6.34**   (1) Difference and (2) reduced IR-LD spectrum of 2-[3-[2-(3,5-dimethoxy-4
-hydroxyvinyl)]-5,5-dimethylcyclohex-2-enylidene]malononitrile after the elimination of the
band at 980 cm⁻¹.

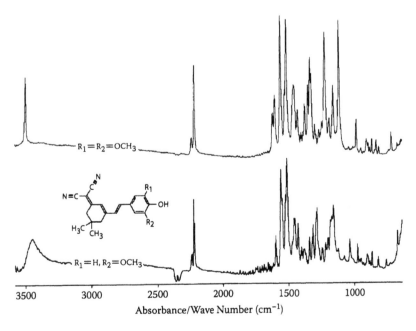

**FIGURE 6.35**   Raman spectra of substituted dicyanoisophorones.

(Figure 6.34). This results in a colinearity of the corresponding transition moments. The theoretically obtained geometry values correlate well with experimental data as far as the bond length differences of less than 0.001 Å. An effectively flat (maximal deviation of 0.8°) geometry of the compound was observed with only one $C(CH_3)_2$ group deviating from the molecule plane at an angle of 22°. The Raman spectra of the methoxy substituted dicyanoisophorones are depicted in Figure 6.35.

Polarized IR-LD spectroscopy of oriented colloids shows a wide application range for the elucidation of the organic dyes in solid state. The CS-NLC method could be successfully applied for elucidation of the π-stacking effects in the aggregates and the tautomerism and stabilization of the quinoide-like form. The method has been combined with corresponding Raman spectroscopic tools and shows excellent correlation between the obtained vibrational data and theoretical conclusions. All of the common rules and advantages could be applied for these systems. One of the most important aspects is that organic dyes are characterized with a significant degree of orientation of suspended particles, allowing precise interpretation of the obtained polarized IR-LD spectra. As far as the used examples are relevant to the field of the design, synthesis, and characterization of new organic materials with nonlinear optical application, the CS-NLC method may also find its place as a modern analytical tool in optics, nonlinear optical technologies, and telecommunications.

# Appendix: List of Acronyms

BLC—baseline correction
BS—Bethe splitting (effect)
CIS—configurational interaction singlet
CBS—common beam spectroscopy
CF—curve fitting
CFP—curve-fitting procedure
CS-NLC—colloid suspensions in a nematic liquid crystal
CT—charge transfer (band)
DAST—4-dimethylamino-N-methyl-4-stilbazolium tosylate
DFT—density functional theory
DS—Davydov splitting (effect)
DSC—differential scanning calorimetry
DTA—differential thermal analysis
EH—Evans's hole (effect)
ESI—electrospray ionization (mass spectrometric method)
FD—Fermi–Davydov (resonance effect)
FFT—fast Fourier transform
FR—Fermi resonance
FSD—Fourier self-deconvolution
FT-IR—Fourier transform infrared
HF—Hartree–Fock
HPLC—high-performance liquid chromatography
HPLC-ESI-MS/MS—high-performance liquid chromatography with tandem mass spectrometry, using electrospray ionization
IR—infrared
IR-LD spectroscopy—linear-dichroic infrared spectroscopy
KBr—potassium bromide
LC—liquid crystal
LOD—limit of detection
MS—mass spectrometry
NLC—nematic liquid crystal
NLO—nonlinear optical
NMR—nuclear magnetic resonance
NPF—nonlinear peak fitting
RDP—reducing-difference procedure
RHF—restricted Hartree–Fock
RS—result spectrum
SbS—subtracted spectrum
SHG—second-harmonic generation
SD—standard deviations
SS—sample spectrum

TD-DFT—time-dependence density functional theory
TGA—thermogravimetric analysis
UHF—unrestricted Hartree–Fock
UV-VIS—ultraviolet-visible spectroscopy
XRD—X-ray diffraction

# References

1. Ambronn, H. 1888. Ueber das optische Verhalten der Cuticula und der verkorkten Membranen. *Ber. Deutsch. Botan. Ges.* 6:85–104.
2. Zbinden, R. 1964. *Infrared Spectroscopy of High Polymers.* Academic Press, New York.
3. Frushour, B.G., P.C. Painter, and J.L. Köning. 1976. Vibrational spectra of polypeptides. *J. Macromol. Sci., Rev. Macromol. Chem. Phys. C* 15:29–115.
4. Jasse, B., and J.L. Köning. 1979. Orientational measurement in polymers, using vibrational spectroscopy. *J. Macromol. Sci., Rev. Macromol. Chem. Phys. C* 17:61–135.
5. Thulstrup, E.W., and J.H. Eggers. 1968. Moment directions of the electronic transitions of fluoranthene. *Chem. Phys. Lett.* 1:690–692.
6. Michl, J., and E.W. Thulstrup. 1986. *Spectroscopy with Polarized Light: Solute Alignment by Photoselection, in Liquid Crystals, Polymers, and Membranes.* VCH, New York.
7. Thulstrup, E.W., and J. Michl. 1989. *Elementary Polarization Spectroscopy.* VCH, New York.
8. Schrader, B. (Ed.). 1995. *Infrared and Raman Spectroscopy.* VCH, Berlin.
9. Siesler, H.W. 1996. Characterization of polymer deformation by vibrational spectroscopy, in: *Orientated Polymer Materials*, St. Fakirov (Ed.). Huethling Wepf Verlag Zug, Oxford.
10. Laurent, M., and B. Samori. 1987. Linear dichroism of solute molecules within micelles. 2. Preferred orientations and local ordering. *J. Am. Chem. Soc.* 109:5109–5113.
11. Samori, B., and L. Fiocco. 1982. Phosphine functionalized macrocycles. A new type of bridging ligand for the synthesis of heterometallic complexes. *J. Am. Chem. Soc.* 104:2634–2637.
12. Gottarelli, G., B. Samori, C. Fuganti, and P. Grasselli. 1981. Liquid crystal characterization of compounds chiral by deuterium substitution. *J. Am. Chem. Soc.* 103:471–472.
13. Vogt, J. 2006. Polarization infrared spectroscopy study of quasi-orthorhombic acetylene thin films on KCl (100). *Phys. Rev. B.* 73:085418–085427.
14. Jordanov, B., E.H. Korte, and B. Schrader. 1988. Differential FTIR spectroscopy with polarized radiation. *J. Mol. Struct.* 174:147–152.
15. Belhakem, M., and B. Jordanov. 1990. FTIR linear dichroic spectra of substances dissolved and oriented in a nematic liquid crystal. *J. Mol. Struct.* 218:309–314.
16. Jordanov, B., R. Nentchovska, and B. Schrader. 1993. FT-IR linear–dichroic solute spectra of nematic solutions as a tool for IR band assignment. *J. Mol. Struct.* 297:401–406.
17. Jordanov, B., R. Nentchovska, and T. Kolev. 1993. Fourier transform infrared polarization spectra of benzaldehyde, terephthaloaldehyde, and terephthalophenone in nematic solutions. *SPIE Proc.* 2089:270–272.
18. Baranovic, G., B. Jordanov, and B. Schrader. 1994. Vibrational states of acenaphthylene: Linear dichroism in nematic solvent and normal coordinate analysis. *J. Mol. Struct.* 323:103–115.
19. Jordanov, B., and B. Schrader. 1995. Reduced IR-LD spectra of substances oriented as nematic solutions. *J. Mol. Struct.* 347:389–398.
20. Andreev, G., E. Korte, B. Jordanov, and B. Schrader. 1997. Polarized FT Raman spectra of trans-stilbene oriented in the nematic liquid crystal host 4-alkyl-4′-cyanobicyclohexyl. *J. Mol. Struct.* 408:305–308.
21. Rogojerov, M., B. Jordanov, and G. Keresztury. 1999. Reduced IR-LD spectra of substances with two conformations oriented in nematic solutions. *J. Mol. Struct.* 480/481:153–160.
22. Thormann, T., M. Rogojerov, B. Jordanov, and E.W. Thulstrup. 1999. Vibrational polarization spectroscopy of fluorene: Alignment in stretched polymers and nematic liquid crystals. *J. Mol. Struct.* 509:93–104.

23. Rogojerov, M., B. Jordanov, and G. Keresztury. 2000. Vibrational analysis of terephthalaldehyde from its IR and Raman spectra in isotropic and anisotropic solutions. *J. Mol. Struct.* 550/551:455–465.

24. Jordanov, B., D. Tsankov, and E.H. Korte. 2003. Peculiarities in the stretching vibrations of the methylene groups. *J. Mol. Struct.* 651/653:101–107.

25. Rogojerov, M., P. Angelova, G. Keresztury, and D. Tsankov. 2007. Vibrational transition moment directions of a terminally p-nitrobenzyl substituted long-chain alkanethiol by polarized infrared spectra and DFT calculations. *Vibr. Spectrosc.* 43:64–70.

26. Rogojerov, M.I., and M.G. Arnaudov. 1994. IR linear dichroic study of high-symmetry molecules dissolved in a liquid crystal solvent. II. Metal hexacarbonyl complexes $M(CO)_6$ (M = Mo, Cr). *J. Mol. Struct.* 321:157–164.

27. Rogojerov, M.I., and M.G. Arnaudov. 1994. IR linear dichroic study of high-symmetry molecules dissolved in a liquid crystal solvent. I. Carbon tetrabromide. *J. Mol. Struct.* 321:147–155.

28. Arnaudov, M.G., B.B. Ivanova, and L.S. Prangova. 2003. Linear-dicroic infrared spectral analysis on the stereo structure of 4'-cyanophenylthiolbenzoate. *J. Mol. Struct.* 661-662:219–226.

29. Arnaudov, M. 2004. The influence of the medium on the infrared spectrum of self-associated systems. *Internet J. Vibr. Spectrosc.* 5(5), sect. 3 (http://www.ijvs.com/volume5/edition5/section3.html).

30. Ivanova, B.B., A. Chapkanov, M. Arnaudov, and I.K. Petkov. 2005. IR-LD spectral study and *ab initio* calculations of 1-phenyl 3-substituted pyrazol-5-ones. *Struct. Chem.* 16:47–53.

31. Arnaudov, M.G., B.B. Ivanova, St.T. Todorov, and S.Io. Zareva. 2006. Reducing-difference infrared spectral analysis of cis- and trans-configurated lactams. *Spectrochim. Acta* 63A:491–500.

32. Ivanova, B.B., M.G. Arnaudov, and P.R. Bontchev. 2004. Linear-dichroic infrared spectral analysis of Cu(I)-homocysteine complex. *Spectrochim. Acta* 60A:855–862.

33. Ivanova, B.B., D.L. Tsalev, and M.G. Arnaudov. 2006. Validation of reducing-difference procedure for the interpretation of non-polarized infrared spectra of n-component solid mixtures. *Talanta* 69:822–828.

34. Ivanova, B.B., V.D. Simeonov, M.G. Arnaudov, and D.L. Tsalev. 2007. Linear-dichroic infrared spectroscopy—Validation and experimental design of the orientation technique as suspension in nematic liquid crystal. *Spectrochim Acta* 67A:66–75.

35. Koleva, B., T. Kolev, V. Simeonov, T. Spassov, and M. Spiteller. 2008. Linear polarized IR-spectroscopy of partial oriented solids as a colloid suspension in nematic liquid crystal—New tool for structural elucidation of the chemical compounds. *J. Incl. Phen.* 61:319–333.

36. Soville, D., W. Russel, and W. Schowaiter. 1989. *Colloidal Dispersions.* Cambridge University Press, Cambridge, UK.

37. Gast, A., and E.M. Zukoski. 1989. Electrorheological fluids as colloidal suspensions. *Adv. Coll. Int. Sci.* 30:153–202.

38. Poulin, P., H. Stark, T.C. Lubensky, and D.A. Weitz. 1997. Novel colloidal interactions in anisotropic fluids. *Science* 275:1770–1773.

39. Poulin, P., N. Frances, and O. Mondain-Monval. 1999. Suspension of spherical particles in nematic solutions of disks and rods. *Phys. Rev. E* 59:4384–4387.

40. Borstnik, A., H. Stark, and S. Zumer. 1999. Interaction of spherical particles dispersed in a liquid crystal above the nematic-isotropic phase transition. *Phys. Rev. E* 60:4210–4218.

41. Lev, B.I., K.M. Aoki, P.M. Tomchuk, and H. Yokoyama. 2003. Structure formation of colloids in nematic liquid crystals. *Condens. Matter Phys.* 6:169–180.

42. Feng, J.J., and C. Zhou. 2004 Orientational defects near colloidal particles in a nematic liquid crystal. *J. Coll. Int. Sci.* 269:72–78.
43. Terentjev, E.M. 1995. Disclination loops, standing alone and around solid particles, in nematic liquid crystals. *Phys. Rev. E* 51:1330–1337.
44. Stark, H. 2001. Physics of colloidal dispersions in nematic liquid crystals. *Phys. Reports* 351:387–474.
45. Loudet, J.-C., P. Barois, and P. Poulin. 2000. Colloidal ordering from phase separation in a liquid-crystalline continuous phase. *Nature London* 407:611–613.
46. Meeker, S.P., W.C.K. Poon, J. Crain, and E.M. Terentjev. 2000. Colloid–liquid-crystal composites: An unusual soft solid. *Phys. Rev. E* 61:R6083.
47. Nazarenko, V.G., A.B. Nych, and B.I. Lev. 2001. Crystal structure in nematic emulsion. *Phys. Rev. Lett.* 87:075504.
48. Stark, H. 2002. Saturn-ring defects around microspheres suspended in nematic liquid crystals: an analogy between confined geometries and magnetic fields. *Phys. Rev. E* 66:032701.
49. Wilson, E.B., J.C. Decius, and P.C. Cross. 1955. *Molecular Vibrations*. McGraw Hill, New York.
50. Hollas, J.M. 2004. *Modern Spectroscopy: Vibrational Spectroscopy*. Wiley, New York.
51. Born, M., and E. Wolf. 1980. *Principles of Optics*. Pergamon Press, Oxford.
52. Maier, W., and A. Saupe. 1959. A simple molecular-statistics theory of the nematic liquid-crystalline state. *Z. Naturforsch.* 13a:564–570.
53. Maier, W., and A. Saupe. 1959. A simple molecular-statistics theory of the nematic liquid-crystalline phase, part I. *Z. Naturforsch.* 14a:882–900.
54. Maier, W., and A. Saupe. 1961. A simple molecular-statistics theory of the nematic liquid-crystalline phase, part II. *Z. Naturforsch.* 16a:287–292.
55. Luckhurst, G.R., and C. Zannoni. 1977. Why is the Maier-Saupe theory of nematic liquid crystals so successful? *Nature* 267:412–414.
56. De Jeu, W.H. 1980. *Physical Properties of Liquid Crystalline Materials*. Gordon and Breach Science, New York.
57. Jaffe, H.H., and M. Orchin. 1965. *Symmetry in Chemistry*. Wiley, New York.
58. Nakamoto, K. 1997. *Infrared and Raman Spectra of Inorganic and Coordination Compounds*. Wiley, New York.
59. Fishman, A.I., A.A. Stolov, and A.B. Remizov. 1993. Vibrational spectroscopic approaches to conformational equilibria and kinetics (in condensed media). *Spectrochim. Acta.* 49A: 1435–1479.
60. Langkilde, F.W., M. Glsin, E.W. Thulstrup, and J. Michl. 1983. Alignment of solutes in stretched polyethylene. Determination of the five second and fourth moments of the orientation distribution of 2-fluoropyrene from polarized fluorescence. Additional evidence for the twisting of weak transition moments by the solvent environment. *J. Phys. Chem.* 87:2901–2911.
61. Spanget-Larsen, J., and N. Fink. 1990. Molecular symmetry of 2,5-dimethyl-1,6,6a. lambda.4-trithiapentalene: Infrared linear dichroism in stretched polyethylene. *J. Phys. Chem.* 94:8423–8425.
62. Spanget-Larsen, J. 1992. Fourier transform infrared linear-dichroism spectroscopy of 1,8-dihydroxy-9,10-anthraquinone aligned in stretched polyethylene. *SPIE Proc.* 1575:404–406.
63. Radziszewski, J.G., J.W. Downing, M.S. Gudipati, V. Balaji, E.W. Thulstrup, and J. Michl. 1996. How predictable are IR transition moment directions? Vibrational transitions in propene and deuterated propenes. *J. Am. Chem. Soc.* 118:10275–10284.
64. Radziszewski, J.G., J. Abilgaard, and E.W. Thulstrup. 1997. The vibrational spectrum of acenaphthylene: linear dichroism and *ab initio* model calculations. *Spectrochim. Acta* 53A:2095–2107.
65. Friedel, G. 1922. Les etats mesomorphes de la matiére. *Ann. Phys. Paris.* 18:273–473.

66. Jordanov, B. 2000. Application of the Linear Dichroism for Analysis of Vibration Transitions, Institute of Organic Chemistry with Centrum of Phytochemistry. DSc diss., Bulgarian Academy of Sciences, Sofia.
67. Korte, E.H., and B. Schrader. 1981. New chiroptical method: infrared rotatory dispersion of induced cholesteric solutions, in: *Advances in Infrared and Raman Spectroscopy*, H. Clark and C. Hester (Eds.). Heyden, London.
68. Jordanov, B., and D. Tsankov. 1982. A differential method for linear dichroic IR spectra—Its theory and applications. *J. Mol. Struct.* 79:5–11.
69. Jordanov, B., and D. Tsankov. 1984. A spectroscopic method for measuring linear birefringence in the IR region—Another application of the common beam technique. *J. Mol. Struct.* 115:457–460.
70. Jordanov, B., and D. Tsankov. 1984. A method for measuring circular dichroism of induced cholesteric solutions in the infrared region. *J. Mol. Struct.* 117:261–264.
71. Ivanova, B.B., M. Arnaudov, W.S. Sheldrick, and H. Mayer-Figge. 2006. S-Phenyl 4-cyanothiobenzoate. *Acta Crystallogr. E* 62:o3–o4.
72. Radziszewski, J.G., and J. Michl. Fourier-transform infrared linear dichroism. Stretched polyethylene as a solvent in IR spectroscopy. *J. Am. Chem. Soc.* 108:3289–3297.
73. Li, K., and S. Banerjee. 1991. Interpreting multicomponent infrared spectra by derivative minimization. *Appl. Spectrosc.* 45:1047–1049.
74. Friese, M.A., and S. Banerjee. 1992. Lignin determination by FT-IR. *Appl. Spectrosc.* 46:246–248.
75. Jordanov, B., M. Rogojerov, and B. Schrader. 1997. Study of overlapped bands by means of reduced IR linear dichroic spectra. *J. Mol. Struct.* 408/409:309–314.
76. Bethe, H. 1929. Termaufspaltung in Kristallen. *Ann. Phys.* 2:133–208.
77. Frenkel, J. 1931. On the transformation of light into heat in solids. II. *Phys. Rev.* 37:1276–1294.
78. Freed, S. 1942. Spectra of ions in fields of various symmetry in crystals and solutions. *Rev. Modern Phys.* 14:105–111.
79. Davydov, A.S. 1962. *Theory of Molecular Excitons*. McGraw Hill, New York.
80. Davydov, A.S. 1968. *Theory of Molecular Excitons*. Nauka, Moscow.
81. Winston, H. 1951. The electronic energy levels of molecular crystals. *J. Chem. Phys.* 19:156–163.
82. McClure, D.S. 1954. Polarization of $\pi$-electron transitions in aromatic molecules. *J. Chem. Phys.* 22:1256–1259.
83. Lisitsa, M.P., N.E. Ralko, and A.M. Yaremko. 1972. Exciton absorption in molecular crystals when Fermi resonance is present. *Phys. Lett. A* 40:329–330.
84. Lisitsa, M., N.E. Ralko, and A.M. Yaremko. 1974. Exciton splitting and Fermi resonance in solid solutions. *Phys. Lett. A* 48:241–243.
85. Lisitsa, M.P., N.E. Ralko, and A.M. Yaremko. 1974. Davydov splitting and Fermi resonance in molecular crystals. *Mol. Cryst. Liq. Cryst.* 28:161–166.
86. Lisitsa, M.P., and A.M. Yaremko. 1972. The temperature dependence of the absorption bands intensities of the spectra of molecular crystals at Fermi resonance I. The relation of the intensities in maxima of the absorption bands. *Mol. Cryst. Liq. Cryst.* 18:297–307.
87. Valakh, M., A. Yaremko, N. Novosele, and M. Lisitsa. 1972. Possibility of Fermi resonance between impurity localized vibration and fundamental lattice vibrations. *Fizika Tverd. Tela.* 14:832–828.
88. Lisitsa, M.P., and A.M. Yaremko. 1972. The temperature dependence of the absorption bands intensities of the spectra of molecular crystals at Fermi resonance. II. The mass operator. The band shape. *Mol. Cryst. Liq. Cryst.* 19:1–11.
89. Marechal, Y., and A. Witkowski. 1968. Infrared spectra of H-bonded systems. *J. Chem. Phys.* 48:3697–3700.

90. Witkowski, A., and Y. Marerchal. 1973. Infrared spectra of hydrogen bond a general theoretical model. *Chem. Phys.* 1:9–16.
91. Wojcik, M. 1978. Fermi resonance in dimers: A model study. *Mol. Phys.* 36:1757–1767.
92. Chamma, D., and O. Henri-Rousseau. 1999. IR theory of weak H-bonds: Davydov coupling, Fermi resonances and direct relaxations. I. Basis equations within the linear response theory. *Chem. Phys.* 248:53–70.
93. Belharaya, K., P. Blaise, and O. Henri-Rousseau. 2003. IR spectral density of weak H-bonds involving indirect damping. I. A new approach using non-Hermitean effective Hamiltonians. *Chem. Phys.* 293:9–22.
94. Bratos, S. 1975. Profiles of hydrogen stretching IR bands of molecules with hydrogen bonds: a stochastic theory. I. Weak and medium strength hydrogen bonds. *J. Chem. Phys.* 63:3499–3501.
95. Bratos, S., and H. Ratajczak. 1982. Profiles of hydrogen stretching IR bands of molecules with hydrogen bonds: A stochastic theory. II. Strong hydrogen bonds. *J. Chem. Phys.* 76:77–79.
96. Ratajczak, H., and A.M. Yaremko. 1995. Theory of profiles of hydrogen stretching infrared bands of hydrogen-bonded solids. Model of strong coupling between the high-frequency hydrogen stretching vibration and low-frequency phonons. *Chem. Phys. Lett.* 243:348–353.
97. Ratajczak, H., and A.M. Yaremko. 1999. Theory of the profiles of hydrogen stretching infrared bands of hydrogen-bonded solids: Fermi resonance effect and strong coupling between the high-frequency hydrogen stretching vibration and low-frequency phonons. *Chem. Phys. Lett.* 314:122–131.
98. Yaremko, A., H. Ratajczak, J. Baran, A. Barnes, E. Mozdor, and B. Silvi. 2004. Theory of profiles of hydrogen bond stretching vibrations: Fermi–Davydov resonances in hydrogen-bonded crystals. *Chem. Phys.* 306:57–70.
99. Witkowski, A. 1967. Infrared spectra of the hydrogen-bonded carboxylic acids. *J. Chem. Phys.* 47:3645–3649.
100. Wojcik, A. 1976. Theory of the infrared spectra of the hydrogen bond in molecular crystals. *Int. J. Quant. Chem.* 10:747–760.
101. Witkowski, A. 1965. In: *Modern Quantum Chemistry*, O. Sinanoglu (Ed.). Academic Press, New York.
102. Witkowski, A., and M. Zcmrski. 1979. Growth rate anisotropy and kinetic coefficients of vicinal faces at LPE garnet films. I. Growth rate anisotropy. *Phys. Stat. Solidi* 46:429–432.
103. Bratos, S., and D. Hards. 1957. Infrared spectra of molecules with hydrogen bonds. *J. Chem. Phys.* 27:991–993.
104. Sheppard, N. 1959. In: *Hydrogen Bonding*, D. Hadzi and H.W. Thompson (Eds.). Pergamon Press, London, p. 85.
105. Claydon, M., and N. Sheppard. 1969. The nature of A,B,C-type infrared spectra of strongly hydrogen-bonded systems; pseudo-maxima in vibrational spectra. *Chem. Comm.* 1969:1431–1433.
106. Evans, J.C. 1960. Further studies of unusual effects in the infrared spectra of certain molecules. *Spectrochim. Acta* 16A:994–1000.
107. Lisitsa, M.P., and A.M. Yaremko. 1970. Combined Fermi-Davydov resonance in solid solutions. *Mol. Cryst. Liq. Cryst.* 6:393–406.
108. Forest, M.G., Q. Hong, and Z.R. Zhou. 2004. Structure scaling properties of confined nematic polymers in plane Couette cells: the weak flow limit. *J. Rheol.* 48:175–192.
109. Poulin, P., V.A. Raghunathan, P. Richetti, and D. Roux. 1994. On the dispersion of latex particles in a nematic solution. I. Experimental evidence and a simple model. *J. Physique II* 4:1557–1569.

110. Andrienko, D., M. Tasinkevych, and S. Dietrich. 2005. Effective pair interactions between colloidal particles at a nematic-isotropic interface. *Europhys. Lett.* 70:95–101.

111. West, J.L., K. Zhang, A. Glushchenko, D. Andrienko, M. Tasinkevych, and Y. Reznikov. 2006. Colloidal particles at a nematic-isotropic interface: Effects of confinement. *Eur. Phys. J. E.* 20:237–242.

112. Smalyukh, I.I., O.D. Lavrentovich, A.N. Kuzmin, A.V. Kachynski, and P.N. Prasad. 2005. Elasticity-mediated self-organization and colloidal interactions of solid spheres with tangential anchoring in a nematic liquid crystal. *Phys. Rev. Lett.* 95:157801.

113. Andrienko, D., M. Tasinkevych, P. Patrício, and M.M. Telo da Gama. 2004. Interaction of colloids with a nematic-isotropic interface. *Phys. Rev. E.* 69:021706.

114. Vollmer, D., G. Hinze, W.C.K. Poon, J. Cleaver, and M.E. Cates. 2004. The origin of network formation in colloid–liquid crystal composites. *J. Phys.: Condensed Matter* 16:L227–L233.

115. Stark, H., J. Fukuda, and H. Yokoyama. 2004. Capillary condensation in liquid-crystal colloids. *Phys. Rev. Lett.* 92:205502.

116. Kalyon, D.M., P. Yaras, B. Aral, and U. Yilmazer. 1993. Rheological behavior of a concentrated suspension: A solid rocket fuel simulant. *J. Rheol.* 37:35–53.

117. West, J.L., A. Glushchenko, G. Liao, Y. Reznikov, D. Andrienko, and M.P. Allen. 2002. Drag on particles in a nematic suspension by a moving nematic-isotropic interface. *Phys. Rev.* E66:012702.

118. Stark, H. 1999. Director field configurations around a spherical particle in a nematic liquid crystal. *Eur. Phys. J. B* 10:311–321.

119. Pusey, P.N., and W. van Megen. 1986. Phase behaviour of concentrated suspensions of nearly hard colloidal spheres. *Nature* 320:340–342.

120. Araki, T., and H. Tanaka. 2006. Colloidal aggregation in a nematic liquid crystal: topological arrest of particles by a single-stroke disclination line. *Phys. Rev. Lett.* 97:127801.

121. Gillette, P.C., and J.L. Koenig. 1984. Objective criteria for absorbance subtraction. *Appl. Spectrosc.* 38:334–337.

122. Savitzky, A., and M.J.E. Golay. 1964. Smoothing and differentiation of data by simplified least squares procedures. *Anal. Chem.* 36:1627–1639.

123. Steiner, J., Y. Termonia, and J. Deltour. 1972. Smoothing and differentiation of data by simplified least square procedure. *Anal. Chem.* 44:1906–1909.

124. Madden, H.M. 1978. Comments on the Savitzky–Golay convolution method for least-squares-fit smoothing and differentiation of digital data. *Anal. Chem. 50*:1383–1586.

125. Barnes, B.J., M.S. Dhanoa, and S.J. Lister. 1989. Standard normal variate transformation and de-trending of near-infrared diffuse reflectance spectra. *Appl. Spectrosc.* 43:772–777.

126. Myers, J., and A. Well. 1991. *Research Design and Statistical Analysis.* Harper Collins, New York.

127. Spiegel, M. 1992. *Theory and Problems of Probability and Statistics.* McGraw-Hill, New York.

128. Baumeister, U., H. Hartung, and M. Jaskolski. 1982. 4′-cyanophenyl-4-pentylbenzoate. *Mol. Cryst. Liq. Cryst.* 88:167–170.

129. Lokanath, N.K., D. Revannasiddaiah, M.A. Sridhar, J.S. Prasad, and D.K. Gowda. 1997. Crystal structure of 4′-cyanophenyl-4-heptylbenzoate. *Zeitschrift für Kristallographie* 212:385–386.

130. Kolev, T.S., B.B. Koleva, and B. Schivachev. 2008. Oriented solids as a colloid suspension in nematic liquid crystal—New tool for IR-spectroscopic and structural elucidation of inorganic compounds and glasses. *Inorg. Chim. Acta* 361:2002–2012.

131. Thomas, P.A. 1988. The crystal structure and absolute optical chirality of paratellurite, α-TeO₂. *J. Phys. C* 21:4611–4627.

132. Mirgorodsky, A.P., T. Merle-Méjean, J.-C. Champarnaud, P. Thomas, and B. Frit. 2000. Dynamics and structure of TeO₂ polymorphs: Model treatment of paratellurite and tellurite; Raman scattering evidence for new γ- and δ-phases. *J. Phys. Chem. Solid.* 61:501–509.

133. Dimitriev, Y., V. Dimitrov, and M. Arnaudov. 1979. Infra-red spectra of crystalline phases and related glasses in the TeO₂-V₂O₅-Me₂O system. *J. Mater. Sci.* 14:723–727.

134. Noruera, O., T. Merle-Mejean, A.P. Mirgorodsky, M.B. Smirnov, P. Thomas, and J.C. Champarnaud-Mesjard. 2003. Vibrational and structural properties of glass and crystalline phases of TeO₂. *J. Non-Crystall. Solid.* 330:50–60.

135. Enjalbert, R., and J. Galy. 1986. A refinement of the structure of V₂O₅. *Acta Crystallogr.* C42:1467–1469.

136. Barroclough, C.G., S. Lewis, and R.S. Nyholm. 1959. The stretching frequencies of metal–oxygen double bonds. *J. Chem. Soc.* 11:3552–3555.

137. Dimitriev, Y., V. Dimitrov, M. Arnaudov, and D. Topalov. 1983. IR-spectral study of vanadate vitreous systems. *J. Non-Crystall. Solid.* 57:147–156.

138. Chen, M., U.V. Waghmare, C.M. Friend, and E. Kaxiras. 1998. A density functional study of clean and hydrogen-covered MoO3(010): electronic structure and surface relaxation. *J. Chem. Phys.* 109:6854–6859.

139. Sian, T., and G.B. Reddy. 2004. Infrared spectroscopic studies on Mg intercalated crystalline MoO₃ thin films. *Appl. Surf. Sci.* 236:1–5.

140. Tokarz-Sobieraj, R., K. Hermann, M. Witko, A. Blume, G. Mestl, and R. Schlögl. 2001. Properties of oxygen sites at the MoO₃(010) surface: Density functional theory cluster studies and photoemission experiments. *Surf. Sci.* 489:107–125.

141. Zhuang, H., X. Zou, and Z. Jin. 1997. Temperature dependence of Raman spectra of vitreous and molten B₂O₃. *Phys. Rev. B* 55:R6105–R6108.

142. Hassan, A.K., L.M. Torell, L. Boerjesson, and H. Doweidar. 1992. Structural changes of B₂O₃ through the liquid–glass transition range: A Raman-scattering study. *Phys. Rev. B* 45:12797–12805.

143. Galeener, F.L., G. Lucovsky, and J.C. Mikkelsen. 1980. Vibrational spectra and the structure of pure vitreous B₂O₃. *Phys. Rev. B* 22:3983–3990.

144. Ardelean, I., N. Muresan, and P. Pascuta. 2006. FT-IR and Raman spectroscopic study of Cr₂O₃-TeO₂-B₂O₃-SrF₂ glasses. *Modern Phys. Lett. B* 20:1107–1114.

145. Barrio, R., F.L. Castillio-Alvarado, and F.L. Galeener. 1991. Structural and vibrational model for vitreous boron oxide. *Phys. Rev. B* 44:7313–7320.

146. Maniu, D., T. Iliescu, I. Andrelean, S. Cinta-Pinzaru, N. Tarcea, and W. Kiefer. 2003. Raman study on B₂O₃-CaO glasses. *J. Mol. Struct.* 651:485–488.

147. Ardelean, I., N. Mureşan, and P. Păşcuţă. 2004. IR and Raman spectroscopixc investigations of Cr₂O₃-TeO₂-B₂O₃-SrO glasses. *Int. J. Modern Phys. B* 18:95–101.

148. Pisarski, W.A., J. Pisarska, and W. Ryba-Romanowski. 2005. Structural role of rare earth ions in lead borate glasses evidenced by infrared spectroscopy: BO₃↔BO₄ conversion. *J. Mol. Struct.* 744:515–520.

149. Moryc, U., and W.S. Ptak. 1999. Infrared spectra of β-BaB₂O₄ and LiB₃O₅: New nonlinear optical materials. *J. Mol. Struct.* 511:241–249.

150. Lu, S.-F., Z.-X. Huang, and J.-L. Huang. 2006. Meta-barium borate, II-BaB₂O₄, at 163 and 293 K. *Acta Crystallogr. C* 62:i73–i75.

151. Willard, C.S., and A.C. Albrecht. 1986. The second-order nonlinear optical properties of β-barium metaborate: A semiempirical theoretical investigation. *Opt. Comm.* 57:146–152.

152. Y. Nakamoto. 1997. *Infrared Spectra of Inorganic and Coordination Compounds.* VCH, New York.

153. Shuvalov, R.R., and P.C. Burns. 2003. A new polytype of orthoboric acid, $H_3BO_3$-3T. *Acta Crystallogr. C* 59:i47–i49.
154. Strong, S.L., A.F. Wells, and R. Kaplow. 1971. On the crystal structure of $B_2O_3$. *Acta Crystallogr. B* 27:1662–1663.
155. Prewitt, C.T., and R.D. Shannon. 1968. Crystal structure of a high-pressure form of $B_2O_3$. *Acta Crystallogr. B* 24:869–874.
156. Dimitriev, Y., J.C.J. Bart, V. Dimitrov, and M. Arnaudov. 1981. Structure of glasses of the $TeO_2$-$MoO_3$ system *Z. Anorg. Allg. Chem.* 479:229–240.
157. Ivanova, B.B., and H. Mayer-Figge. 2005. Crystal structure and solid state IR-LD spectral analysis of new mononuclear Cu(II) complex with 4-aminopyridine. *J. Coord. Chem.* 58:653–659.
158. Ivanova, B.B., M.G. Arnaudov, and H. Mayer-Figge. 2005. Molecular spectral analysis and crystal structure of 4-aminopyridinium tetrachloropalladate(II) complex salt. *Polyhedron* 24:1624–1630.
159. Koleva, B.B., E.N. Trendafilova, M.G. Arnaudov, W.S. Sheldrick, and H. Mayer-Figge. 2006. Structural analysis of a mononuclear copper(II) complex of 3-aminopyridine. *Trans. Met. Chem.* 31:866–873.
160. Kolev, T., B.B. Koleva, T. Spassov, E. Cherneva, M. Spiteller, W.S. Sheldrick, and H. Mayer-Figge. 2008. Synthesis, spectroscopic, thermal and structural elucidation of 5-amino-2-methoxypyridine ester amide of squaric acid ethyl ester: A new material with an infinite pseudo-layered structure and manifested NLO application. *J. Mol. Struct.* 875:372–381.
161. Koleva, B., T. Tsanev, T. Kolev, H. Mayer-Figge, and W. S. Sheldrick. 2007. 3,4-diaminopyridinium hydrogensquarate. *Acta Crystallogr. E* 63:o3356–o3358.
162. Koleva, B.B., T. Kolev, T. Tsanev, St. Kotov, H. Mayer-Figge, R.W. Seidel, and W.S. Sheldrick. 2008. Crystal structure, optical and magnetic properties of the bis(perchlorate) of 3,4-diaminopyridine. *Struct. Chem.* 19:13–20.
163. Koleva, B.B., T. Kolev, T. Tsanev, St. Kotov, H. Mayer-Figge, R.W. Seidel, and W.S. Sheldrick. 2008. Spectroscopic and structural elucidation of 3,4-diaminopyridine and its hydrogentartarate salt: Crystal structure of 3,4-diaminopyridinium hydrogentartarate dihydrate. *J. Mol. Struct.* 881:146–155.
164. Ivanova, B.B. 2005. N1 protonated salt of adenine—Solid-state linear dichroic infrared spectral analysis. *Spectrosc. Lett.* 38:635–643.
165. Kolev, T., B.B. Koleva, M. Spiteller, W.S. Sheldrick, and H. Mayer-Figge. 2007. 2-amino-4-nitroaniline, a known compound with unexpected properties. *J. Phys. Chem.* 111A:10084–10089.
166. Ivanova, B.B. 2005. Solid state linear dichroic infrared spectral analysis of benzimidazoles and their N1-protonated salts. *Spectrochim. Acta* 62A:58–65.
167. Ivanova, B.B., and L.I. Pindeva. 2006. Protonation of benzimidazoles and 1,2,3-benzotriazoles—Solid state linear dichroic infrared (IR-LD) spectral analysis and *ab intio* calculations. *J. Mol. Struct.* 797:144–153.
168. Ivanova, B.B., and L.I. Pindeva. 2006. IR-LD spectroscopy of benzyl oriented as solution in nematic liquid crystal and melted polycrystalline solid sample. *Bull. Chemist. Tech. Mac.* 25:39–43.
169. Ivanova, B.B., and M. Mitewa. 2004. Au(III) interaction with methionine and histidine containing peptides. *J. Coord. Chem.* 57:217–222.
170. Ivanova, B.B., and H. Mayer-Figge. 2004. Crystal structure of zwitterionic β-chloro-L-alanine. *Anal. Sciences* 20:x73–x75.
171. Ivanova, B.B. 2005. Solid state linear-dichroic infrared spectral analysis of dipeptide L-Phe-L-Phe and its mononuclear Au(III)-complex. *J. Coord. Chem.* 58:587–593.
172. Ivanova, B.B. 2006. Solid state linear-dichroic infrared (IR-LD) spectral characterization of α- and β-polymorphs of glycine. *Centrl. Europ. J. Chem.* 4:111–117.

173. Ivanova, B.B., and M.G. Arnaudov. 2006. Theoretical and solid state linear-dichroic infrared spectral analysis of aromatic dipeptides and their protonated forms. *Protein Pep. Lett.* 13:889–896.

174. Ivanova, B.B., M.G. Arnaudov, and St.T. Todorov. 2006. Linear-dichroic infrared and NMR spectral analysis of Au³⁺-complex with the tripeptide glycyl-methionyl-glycine. *J. Coord. Chem.* 59:1749–1755.

175. Ivanova, B.B. 2006. Stereo-structural and IR-spectral characterization of histidine containing dipeptides by means of solid-state IR-LD spectroscopy and *ab initio* calculations. *J. Mol. Struct.* 782:122–129.

176. Ivanova, B.B., and M.G. Arnaudov. 2006. Solid state linear-dichroic infrared spectral and theoretical analysis of methionine containing tripeptides. *Spectrochim. Acta* 65A:56–61.

177. Ivanova, B.B., M.G. Arnaudov, St. Todorov, W.S. Sheldrick, and H. Mayer-Figge. 2006. Structural analysis of the tripeptide glycyl-methionyl-glycine (H-Gly-Met-Gly-OH) and its hydrochloride. *Struct. Chem.* 17:49–56.

178. Ivanova, B.B. 2006. IR-LD spectroscopic characterization of L-tryptophan containing dipeptides. *Spectrochim. Acta* 64A:931–938.

179. Ivanova, B.B. 2008. Aromatic dipeptides and their salts—Solid-state linear-dichroic infrared (IR-LD) spectral analysis and *ab initio* calculations. *Spectrochim. Acta* 70A:324–331.

180. Ivanova, B.B., T. Kolev, and S.Y. Zareva. 2006. Solid-state IR-LD spectroscopic and theoretical analysis of glycine-containing peptides and their hydrochlorides. *Biopolymers* 82:587–596.

181. Kolev, T., B.B. Ivanova, E. Cherneva, M. Spiteller, W. S. Sheldrick, and H. Mayer-Figge. 2006. Crystal structure, IR-LD spectroscopic, theoretical and vibrational analysis of valinamide ester amide of squaric acid. *Struct. Chem.* 17:491–499.

182. Kolev, T., B.B. Ivanova, and S.Y. Zareva. 2007. Au(III)-complex with dipeptide glycyl-serine—Linear polarized IR-spectroscopic, ¹H and ¹³C magnetic resonance characterization. *J. Coord. Chem.* 60:109–115.

183. Koleva, B.B., T.S. Kolev, S.Y. Zareva, and M. Spiteller. 2007. The dipeptide alanylphenylalanine (H-Ala-Phe-OH)—Protonation and coordination ability with Au(III). *J. Mol. Struct.* 831:165–173.

184. Koleva, B.B., T. Kolev, S.Y. Zareva, and M. Spiteller. 2008. Synthesis, spectroscopic and theoretical characterization of hydrogensquarate and mononuclear Au(III)-complex of dipeptide phenylalanyltyrosine. *J. Mol. Struct.* 855:104–110.

185. Kolev, T., B.B. Koleva, and M. Spiteller. 2007. Spectroscopic and theoretical characterization of hydrogensquarates of L-threonyl-L-serine and L-serine. Prediction of structures of the neutral and protonated forms of the dipeptide. *Amino Acids* 33:719–725.

186. Koleva, B.B. 2007. Solid-state linear-polarized IR-spectroscopic characterization of L-methionine. *Vibr. Spectroscopy* 44:30–35.

187. Kolev, T.S., S.Y. Zareva, B.B. Koleva, and M. Spiteller. 2006. Au(III) complexes of di- and tripeptide glycylalanine and glycylalanylalanine. *Inorg. Chim. Acta.* 359:4367–4376.

188. Koleva, B.B., T.M. Kolev, and M. Spiteller. 2008. Solid-state linear-polarized spectroscopy of alanyl-containing peptides and their hydrogensquarates. *J. Mol. Struct.* 877:79–88.

189. Koleva, B.B., T.M. Kolev, and M. Spiteller. 2006. Structural and spectroscopic analysis of hydrogensquarates of glycine-containing tripeptides. *Biopolymers* 83:498–507.

190. Koleva, B.B., Ts. Kolev, and M. Spiteller. 2007. Mononuclear Au(III)-complexes with tryptophan-containing dipeptides: Synthesis, spectroscopic and structural elucidation. *Inorg. Chim. Acta* 360:2224–2230.

191. Kolev, T.M., B.B. Koleva, M. Spiteller, W.S. Sheldrick, and H. Mayer-Figge. 2009. Synthesis, spectroscopic and structural elucidation of tyrosinamide hydrogensquarate monohydrate. *Amino Acids* 36:195–201.

192. Kolev, T.M., B.B. Koleva, M. Spiteller, W.S. Sheldrick, and H. Mayer-Figge. 2009. Synthesis, spectroscopic and structural elucidation of sympathomimetic amine, tyraminium dihydrogenphosphate. *Amino Acids* 36:185–193.

193. Koleva, B.B., T.M. Kolev, and St. Todorov. 2007. Structural and spectroscopic analysis of dipeptide methionyl-glycine and its hydrochloride. *Chem. Papers* 61:490–496.

194. Koleva, B.B. and T.M. Kolev, and M. Spiteller. 2007. Spectroscopic and structural elucidation of L-tyrosine-containing dipeptides valyl-tyrosine and tyrosyl-alanine: Solid-state IR-LD spectroscopy, quantum chemical calculations and vibrational analysis. *Spectrochim. Acta* 68A:1187–1196.

195. Koleva, B.B., T. Kolev, S. Zareva, M. Lamshöft, and M. Spiteller. 2008. Synthesis, spectroscopic and structural elucidation of Au(III), Pd(II) and Pt(II) complexes with tripeptide glycyl-L-phenylalanyl-glycine. *Trans. Met. Chem.* 33:911–919.

196. Koleva, B.B., T. Kolev, R.W. Seidel, H. Mayer-Figge, and W.S. Sheldrick. 2008. Bis(tyrammonium) sulfate dihidrate: Crystal structure, solid state IR-spectroscopic and theoretical characterization. *J. Mol. Struct.* 888:138–144.

197. Kolev, B.B., T. Kolev, H. Mayer-Figge, R.W. Seidel, and W.S. Sheldrick. 2008. Are there preferable conformations of the tryptammonium cation in the solid state? Crystal structure and solid-state linear polarized IR-spectroscopic study of tryptammonium hydrogentartarate. *Struct. Chem.* 19:147–154.

198. Kolev, T., B.B. Koleva, and M. Spiteller. 2008. New copper(II) complexes with hydroxyl containing dipeptides glycyl-L-serine and L-seryl-L-tyrosine. *J. Coord. Chem.* 61:1897–1905.

199. Koleva, B.B., T. Kolev, S. Zareva, and M. Spiteller. 2008. New Au(III), Pt(II) and Pd(II) complexes with glycyl-containing homopeptides. *J. Coord. Chem.* 61:3534–3548.

200. Kolev, T., B. Ivanova, I. Stoineva, E. Bratovanova, and B. Tchorbanov. 2006. A stable conformer of proline tripeptides with ACE-inhibitory activity proved by solid-state IR-LD spectroscopic and theoretical analysis. *J. Pept. Sci.* 12:133–133 Suppl.

201. Kolev, T., B.B. Koleva, R.W. Seidel, H. Mayer-Figge, M. Spiteller, and W.S. Sheldrick. 2009. Tyrammonium 4-nitrophthalate dihydrate—Structural and spectroscopic elucidation. *Amino Acids* 36:29–33.

202. Kolev, T., R.W. Seidel, M. Spiteller, H. Mayer-Figge, W.S. Sheldrick, and B.B. Koleva. 2009. New structural motifs and nonlinear optical properties of the crystals of squaric acid in the presence of L-lysinium counterion. *J. Mol. Struct.* 919:246–254.

203. Koleva, B.B. 2008. IR-polarized spectroscopic and structural elucidation of protonated tripeptide glycyl-L-phenylalanyl-glycine. *Prot. Pept. Lett.* 15:309–313.

204. Koleva, B. 2008. Solid-state IR-LD spectroscopy of L-tryptophan-containing dipeptides L-tryptophyl-L-methionine (H-Trp-Met-OH), L-methionyl-L-tryptophan (H-Met-Trp-OH) and glycyl-L-tryptophan dehydrate (H-Gly-Trp-OH·2H$_2$O). *Bulg. Chem. Commun.* 40:456–463.

205. Koleva, B.B., T.M. Kolev, D.L. Tsalev, and M. Spiteller. 2008. Determination of phenacetin and salophen analgetics in solid binary mixtures with caffeine by infrared linear dichroic and Raman spectroscopy. *J. Pharmaceut. Biomed. Analys.* 46:267–273.

206. Koleva, B.B. 2006. Polymorphs of Aspirin—Solid-state IR-LD spectroscopic and quantitative determination in solid mixtures. *J. Mol. Struct.* 800:23–27.

207. Ivanova, B.B. 2005. Monoclinic and orthorhombic polymorphs of paracetamol—Solid state linear dichroic infrared spectral analysis. *J. Mol. Struct.* 738:233–238.

208. Koleva, B.B., T.M. Kolev, and M. Spiteller. 2008. Determination of cephalosporins in solid binary mixtures by polarized IR- and Raman spectroscopy. *J. Pharm. Biomed. Anal.* 48:201–204.

209. Ivanova, B.B., T. Kolev, and R. Bakalska. 2007. Linear-dichroic infrared spectral (IR-LD) analysis of codeine and its derivatives. *Spectrochim. Acta* 67:196–201.

210. Kolev, T., R. Bakalska, R.W. Seidel, H. Mayer-Figge, I.M. Oppel, M. Spiteller, W.S. Sheldrick, and B.B. Koleva. 2009. Novel codeinone derivatives via Michael addition of barbituric acids. *Tetrahedron Asymm.* 20:327–334.

211. Kolev, T., B.B. Ivanova, and R. Bakalska. 2006. 6-O-Acetylcodeine and its hydrogensquarate—Linear-dichroic infrared (IR-LD) spectroscopy. *J. Mol. Struct.* 794:138–141.

212. Bakalska, R., B.B. Ivanova, and T.S. Kolev. 2006. Solid-state IR-LD spectroscopy of codeine and N-norcodeine derivatives. *Centrl. Europ. J. Chem.* 4:533–542.

213. Ivanova, B.B. 2006. Solid-state linear-dichroic IR-spectroscopy of isophorone derivatives with potential non-linear optical application. *Spectrochim. Acta* 65A:1035–1040.

214. Koleva, B.B., and T. Kolev. 2008. Solid-state IR-LD spectroscopy of isophorones with potential non-linear optical application: monoclinic and triclinic polymorphs of 2-{5,5-dimethyl-3-[2-(2,4,6-trimethoxyphenyl)vinyl]cyclohex-2-enylidene}malononitrile. *Spectrochim. Acta* 71A: 786–793.

215. Kolev, T.S., B.B. Koleva, M. Spiteller, H. Mayer-Figge, and W.S. Sheldrick. 2008. Synthesis, spectroscopic and structural characterization of 1-methyl-4-[2-(4-hydroxyphenyl)ethenyl)]pyridinium] dihydrogenphosphate: New aspects on crystallographic disorder and its effect on polarized solid-state IR spectra. *Dyes Pigm.* 79:7–13.

216. Kolev, T., B.B. Koleva, S.T. Stoyanov, M. Spiteller, and I. Petkov. 2008. The aggregation of the merocyanine dyes, depending of the type of the counter ions. *Spectrochim. Acta* 70A:1087–1096.

217. Koleva, B.B., S. Stoyanov, T. Kolev, I. Petkov, and M. Spiteller. 2008. Spectroscopic and structural elucidation of merocyanine dye 2,5-[1-metyl-4-[2-(4-hydroxyphenyl)ethenyl)]piridinium]-hexane tetraphenylborate. Aggregation processes. *Spectrochim. Acta* 71A:847–853.

218. Stoyanov, S., B.B. Koleva, T. Kolev, I. Petkov, and M. Spiteller. 2008. Structural elucidation, optical and magnetic properties of tetraphenylborate salt of 2,5-[1-metyl-4-[2-(4-hydroxyphenyl)ethenyl)]piridinium]-butane. *Polish J. Chem.* 82:2167–2178.

219. Koleva, B., S.T. Stoyanov, T. Kolev, I. Petkov, and M. Spiteller. 2009. Structural elucidation, optical, magnetic and nonlinear optical properties of oxystyryl dyes. *Spectrochim. Acta.* 71A:1857–1864.

220. Koleva, B. 2008. Aromatic peptides—Modern tools for spectroscopic and structural elucidation, *Bulg. Chem. Industry* 79:1–6.

221. Kolev, T., B.B. Koleva, M. Spiteller, W.S. Sheldrick, and H. Mayer-Figge. 2008. Benzamidinium D-glucuronate: spectroscopic and structural elucidation. *J. Mol. Struct.* 879:30–39.

222. Kolev, T., R.W. Seidel, B.B. Koleva, M. Spiteller, H. Mayer-Figge, and W.S. Sheldrick. 2008. Crystal structure and spectroscopic properties of ammonium hydrogen squarate squaric acid monohydrate. *Struct. Chem.* 19:101–107.

223. Koleva, B.B., Ts. Kolev, R.W. Seidel, H.Mayer-Figge, M. Spiteller, and W.S. Sheldrick. 2008. On the origin of the colour in the solid-state. *J. Phys. Chem.* 112A:2899–2905.

224. Kolev, T. 2007. Quantum chemical, spectroscopic and structural study of hydrochlorides, hydrogensquarates and ester amides of squaric acid of amino acid amides, in: *New Approaches in Quantum Chemistry*, E. O. Hoffman (Ed.). Nova Science, New York.

225. Koleva, B.B., T. Kolev, and M. Spiteller. 2008. Spectroscopic and structural elucidation of small peptides, in: *Computational Chemistry: New Research*, Nova Science, New York.

226. Kolev, T., M. Spiteller, and Koleva, B. 2010. Spectroscopic and structural elucidation of amino acid derivatives and small peptides—Experimental and theoretical tools. *Amino Acids, Review.* 38:45–50.

227. Koleva, B. B., T. Kolev, R. W. Seidel, M. Spiteller, H. Mayer-Figge, and W. S. Sheldrick. 2009. Self-assembly of hydrogensquarates: Crystal structures and properties. *J. Phys. Chem.* 113A:3088–3095.

228. Kolev, T., B. B. Koleva, R.W. Seidel, M. Spiteller and W.S. Sheldrick. 2009. New aspects on the origin of color in the solid state. Coherently shifting of the protons in violurate crystals. *Cryst. Growth Des.* 9:3348–3352.

229. Kirsch, G.E., and T. Narahashi. 1978. 3,4-diaminopyridine. A potent new potassium channel blocker. *Biophys. J.* 22:507–512.

230. McEvoy, K.M., A.J. Windebank, J.R. Daube, and P.A. Low. 1989. 3,4-Diaminopyridine in the treatment of Lambert–Eaton myasthenic syndrome. *New Eng. J. Med.* 321:1567–1581.

231. Raust, J.A., S. Goulay-Dufay, M.D. Le Hoang, D. Pradeau, F. Guyon, and B. Do. 2007. Stability studies of ionised and non-ionised 3,4-diaminopyridine: Hypothesis of degradation pathways and chemical structure of degradation products. *J. Pharmaceut. Biomed. Anal.* 43:83–88.

232. Sahili, S., B. Frank, B. Schweizer, F. Diderich, D. Kaelin, J. Aebi, H. Bohm, H. Oefner, and G. Dale. 2005. Second-generation inhibitors for the metalloprotease neprilysin based on bicyclic heteroaromatic scaffolds: synthesis, biological activity, and X-ray crystal-structure analysis. *Helv. Chim. Acta* 88:731–750.

233. Niño, A., and C. Muñoz-Caro. 2001. Theoretical analysis of the molecular determinants responsible for the $K^+$ channel blocking by aminopyridines. *Biophys. Chem.* 91:49–60.

234. Rubin-Preminger, J.M., and U. Englert. 2007. 3,4-Diaminopyridine. *Acta Crystallogr. E* 63:o757–o758.

235. Gong, Y., C. Hu, H. Li, W. Tang, K. Huang, and W. Hou. 2006. Synthesis, crystal structure and calcination of three novel complexes based on 2-aminopyridine and polyoxometalates. *J. Mol. Struct.* 784:228–238.

236. Willett, R., H. Place, and M. Middleton. 1988. Crystal structures of three new copper(II) halide layered perovskites: Structural, crystallographic, and magnetic correlations. *J. Am. Chem. Soc.* 110: 8639–8650.

237. Rubin-Preminger, J.M., and U. Englert. 2007. 3,4-Diaminopyridine. *Acta Crystallogr. E* 63:o757–o758.

238. Vembu, N., N. Nallu, J. Garrison, and W. Youngs. 2003. 4-Dimethylaminopyridinium picrate: supramolecular aggregation through extensive N-HO and C-HO interactions. *Acta Crystallogr. E* 59:o913–o916.

239. Nassimbeni, L.R., A. Rodgers, and J. Haigh. 1976. The crystal and molecular structure of the bis(4-N,N′-dimethylaminopyridine) solvate of di-μ-salicylicacidato bis{nitratodioxouranium(VI). *Inorg. Chim. Acta* 20:149–153.

240. Vembu, N., M. Nallu, E. Spencer, and J. Howard. 2003. 4-Dimethylaminopyridinium 2,4-dinitrophenolate: Supramolecular aggregation through N-HO, C-HO, C-H and interactions. *Acta Crystallogr. E* 59:o1383–o1386.

241. Brzozowski, Z., F. Saczewski, and M. Gdaniec. 2003. Synthesis, structural characterization and *in vitro* antitumor activity of 4-dimethylaminopyridinium (6-chloro-1,1-dioxo-1,4,2-benzodithiazin-3-yl)methanides. *Eur. J. Med. Chem.* 38:991–999.

242. Mayr-Stein, R., and M. Bolte. 2000. 4-Dimethylaminopyridinium bromide. *Acta Crystallogr. C* 56:e19–e20.

243. Sluka, R., M. Necas, and M. Cernik. 2003. Bis(4-dimethylaminopyridinium) 1,2-difluorodiphosphate. *Acta Crystallogr. E* 59:o190–o192.

244. Bryant Jr., G.L., and J. King Junior. 1992. Structures of two acylpyridinium salts and one simple pyridinium salt. *Acta Crystallogr. C* 48:2036–2039.

245. Hosomi, H., S. Ohba, and Y. Ito. 2000. CT complexes of 4-(dimethylamino)pyridine with 2,4-, 3,4- and 3,5-dinitrobenzoic acid. *Acta Crystallogr. E* 56:e149–e150.

246. Dega-Szafran, Z., M. Gdaniec, M. Grundwald-Wyspianska, Z. Kosturkiewicz, J. Koput, P. Krzyzanowski, and M. Szafran. 1992. X-ray, FT-IR and PM3 studies of hydrogen bonds in complexes of some pyridines with trifluoroacetic acid. *J. Mol. Struct.* 270:99–124.
247. Porcs-Makkay, M., G. Argay, A. Kalman, and G. Simig. 2000. Synthesis of 1,3-Di[alkoxy(aryloxy)carbonyl]-2-oxo-2,3-dihydroindoles. *Tetrahedron* 56:5893–5903.
248. Porcs-Makkay, M., and G. Simig. 2000. New practical synthesis of tenidap. *Org. Progr. Res. Develop.* 4:10–16.
249. Kin-Shan, H., D. Britton, M. Etter, and S. Byrn. 1997. A novel class of phenol–pyridine co-crystals for second harmonic generation. *J. Mater. Chem.* 7:713–721.
250. Haddad, S., and R. Willett. 2001. Crystal structure of bis(4-dimethylaminopyridinium) tribromocuprate(I) and bis(4-dimethylaminopyridinium) tetrabromocuprate(II). *J. Chem. Crystallogr.* 31:37–43.
251. Haddad, S., and R. Willett. 2001. Polymorphism in bis(4-dimethylamino-pyridinium) tetrachlorocuprate(II). *Inorg. Chem.* 40:2457–2460.
252. Yano, S., M. Kato, K. Tsukahara, M. Sato, T. Shibahara, K. Lee, Y. Sugiara, M. Iida, K. Goto, S. Aoki, Y. Shibahara, H. Nakahra, H. Yanagisawa, and H. Miyagawa. 1994. Chemical modification of metal complexes. An efficient procedure for o-acylation of an anionic metal complex having a noncoordinated hydroxyl group. *Inorg. Chem.* 33:5030–5035.
253. Majerz, I., Z. Malarski, and W. Sawka-Dobrowolska.1995. Structure and IR spectroscopic properties of the anhydrous 4-N,N-dimethylaminopyridinium pentachlorophenolate. *J. Chem. Crystallogr.* 25:189–193.
254. Smith, G., U.D. Wermuth, P.C. Healy, and J.M. White. 2003. Structure-making with 3,5-dinitrosalicylic acid. II. The proton-transfer compounds of 3,5-dinitrosalicylic acid with the monocyclic heteroaromatic amines. *Austr. J. Chem.* 56:707–713.
255. Ren, P., J. Qin, T. Liu, and S. Zhang. 2004. Synthesis, structure and second harmonic generation of novel inorganic–organic hybrid, (p-cyano-1-hydrogenpyridinium)$_2$CdI$_4$. *Inorg. Chem. Commun.* 7:134–136.
256. Wilson, E. B. 1934. The normal modes and frequencies of vibration of the regular plane hexagon model of the benzene molecule. *Phys. Rev.* 45:706–714.
257. Spiner, E. 1962. The vibration spectra and structures of the hydrochlorides of aminopyridines. *J. Chem. Soc.* 3119–3127.
258. Batts, B.D., and E. Spiner. 1969. Vibration-spectral and structural comparison of the 4-aminopyridine cation with the 4-hydroxypyridinium ion and 4-pyridone. The protio parent ions, N- and C-deuterated, and N-methylated ions. Relevant N.M.R. spectral studies. *Austr. J. Chem.* 22:2595–2610.
259. Laing, M., N. Sparrow, and P. Sommerville. 1971. The crystal structure of 4-cyanopyridine. *Acta Crystallogr. B* 27:1986–1990.
260. Ranninger, M., S. Carrera, and S. Blanco. 1985. Crystal and molecular structure of the tetrachloropalladate of the dication meso-3,7-diazonia tricyclo[4.2.2.22,5]dodeca-3,7,9,11-tetraen-4,8 diamine obtained by coupling reaction of 2-aminopyridines. *Polyhedron* 4:1379–1381.
261. Sugiyama, J., J. Meng, and T. Matsurra. 2002. Two-component molecular crystals composed of chloronitrobenzoic acids and 4-aminopyridine. *Acta Crystallogr. C* 58:o242–o246.
262. Mishina, S., M. Takayanagi, M. Nakata, J. Otsuki, and K. Araki. 2001. Dual fluorescence of 4-dimethylaminopyridine and its derivatives: Effects of methyl substitution at the pyridine ring. *J. Photochem. Photobiol. A: Chem.* 141:153–158.
263. Yamada, S., K. Yamaguchi, and K. Ohta. 2002. Theoretical study on second hyperpolarizability for cationic pyridine derivatives. *Mol. Phys.* 100:1839–1846.
264. Hamed, K., A. Samah, and R. Mohamed. 2007. 3-Amino-2-chloropyridinium dihydrogenphosphate. *Acta Crystallogr.* 63:o2896–o2897.

265. Ishida, H. 2004. 1,2,3,4-Tetrahydroquinolinium hydrogen chloranilate. *Acta Crystallogr.* 60:o1674–o1675.
266. Smith, G., U.D. Wermuth, and J.M.White. 2002. The 1:1 adduct of 1,3,5-trinitrobenzene with 1,2,3,4-tetrahydroquinoline. *Acta Crystallogr. E* 58:o1130–o1132.
267. Smith, G., U.D. Wermuth, P.C. Healy, and J.M. White. 2007. 3,5-Dinitrosalicylic acid in molecular assembly. III. Proton-transfer compounds of 3,5-dinitrosalicylic acid with polycyclic aromatic and heteroaromatic amines, and overall series structural systematics. *Austr. J. Chem.* 60:264–277.
268. De Vries, E.J.C., C.L. Oliver, and G.O. Lloyd. 2005. (4-Pyridylmethyl)aminium chloride. *Acta Crystallogr. E* 61:o1577–o1578.
269. Kolev, T., and T. Tsanev. 2008. Spectroscopic elucidation of hydrogensquarate and ester amide of squaric acid of 2-cloro-3-amino-pyridine. *Bulg. Chem. Commun.* 40:483–490.
270. Arnaudov, M.G., B.B. Ivanova, and Sh.G. Dinkov. 2005. A solid-state linear dichroic infrared spectral study of 4-aminopyridine. *Vibr. Spectrosc.* 37:145–147.
271. Saha, B., A. Nangia, and J. Nicoud. 2006. Using halogen-halogen interactions to direct noncentrosymmetric crystal packing in dipolar organic molecules. *Cryst. Growth Des.* 6:1278–1281.
272. Arnaudov, M. 2005. Solid-state linear-dichroic infrared (IR-LD) spectral analysis of samples oriented as a suspension in a nematic liquid crystal. *Bulg. Chem. Commun.* 37:230–237.
273. Arnaudov, M., and B. Ivanova. The effectiveness of the reducing-difference procedure as a tool for an IR-LD spectra interpretation of solids. *Bul. Chem. Commun.* 37:283–288.
274. Koleva, B.B., T. Kolev, R. Nikolova, Y. Zagraniarsky, and M. Spiteller. 2008. Newly synthesized organic material with potential NLO application—Electronic and spectroscopic properties. *Centr. Europ. J. Chem.* 6:592–599.
275. Kolev, T., B. Koleva, R. Sedel, M. Spiteller, and W.S. Sheldrick. 2007. Cyclohexylammonium hydrogensquarate semihydrate. *Acta Crystallogr. E* 63:o4852–o4855.
276. Angelova, O., V. Velikova, T. Kolev, and V. Radomirska. 1996. Crystalline complexes involving amino acids. I. L-argininium hydrogen squarate. *Acta Crystallogr. C* 52:3252–3256.
277. Kolev, T., R. Stahl, H. Preut, P. Bleckmann, V. Radomirska. 1998. Crystal structure of L-(+)-serinium hydrogensquarate, $C_7H_9NO_7$. *Zeitschrift für Kristallographie* 213:169–170.
278. Kolev, T., H. Preut, P. Bleckmann, and V. Radomirska. 1997. Guanidinium hydrogen squarate. *Acta Crystallogr. C* 53:805–807.
279. Cherneva, E., and T. Kolev. 2008. Solid-state polarized IR-spectroscopic study of high temperature red phase of 5-amino-2-methoxypyridine ester amide of squaric acid ethyl ester. *Bulg. Chem. Commun.* 40: 477–482.
280. Zyss, J., and D. Chemla. 1987. Quadratic nonlinear optics, in: *Nonlinear Optical Properties of Organic Molecules and Crystals*, D. Chemla and J. Zyss (Eds.). Academic Press, New York.
281. Nalwa, H.S., T. Watanabe, and S. Miyata. 1999. Organic materials for second-order nonlinear optics, in: *Nonlinear Optics of Organic Molecules and Polymers*, H.S. Nalwa and S. Miyata (Eds.). CRC Press, Boca Raton.
282. Silva, C.E., R. Diniz, B. L. Rodrigues, and L.F.C. de Oliveira. 2007. Crystal structure and spectroscopic analysis of the asymmetric squaraine [(2-dimethylamino-4-anilino) squaraine]. *J. Mol. Struct.* 831:187–194.
283. Spassova, M., T. Kolev, and I. Kanev. 2000. Structure and nonlinear electrical properties of squaric acid derivatives: a theoretical study of the conformation and deprotonation effects. *J. Mol. Struct.: THEOCHEM* 528:151–159.

284. Torii, H., and M. Tasumi. 1995. Vibrational analysis of the squarate ion based on *ab initio* molecular orbital calculations. A practical method to calculate vibrational force fields of non-bond-alternating conjugated molecules. *J. Mol. Struct.: THEOCHEM* 334:15–27.

285. Marder, S. R., J. W. Perry, and W. P. Schaefer. 1989. Synthesis of organic salts with large second-order optical nonlinearities. *Science* 245:626–628.

286. Suwan, S., M. Isobe, O. Yamashita, H. Minakata, and K. Imai. 1994. Silkworm diapause hormone, structure–activity relationships indispensable role of C-terminal amide. *Insect Biochem. Mol. Biology* 24:1001–1007.

287. Kolev, T., E. Cherneva, M. Spiteller, W.S. Sheldrick, and H. Mayer-Figge. 2006. L-Methioninamide ester amide of squaric acid diethyl ester. *Acta Crystallogr. E* 62:o1390–O1392.

288. Kolev, T., B. Bucholz, and M. Spiteller. 2007. Private communication.

289. Kolev, T., D. Yancheva, M. Spiteller, W.S. Sheldrick, and H. Mayer-Figge. 2006. L-Prolinamidium hydrogensquarate. *Acta Crystallogr. E* 62:o463–o465.

290. Kolev, T., M. Spiteller, W.S. Sheldrick, and H. Mayer-Figge. 2005. 1-(Aminocarbonyl) ethylammonium hydrogensquarate monohydrate. *Acta Crystallogr. E* 61:o4292–o4294.

291. Kolev, T., M. Spiteller, W.S. Sheldrick, and H. Mayer-Figge. 2006. L-Arginineamide bis(hydrogensquarate. *Acta Crystallogr. C* 62:o299–o300.

292. Kolev, T., B. Bucholz, and M. Spiteller. 2007. Private communication.

293. Head-Gordon, T., M. Head-Gordon, M. Frisch, C. Brooks, and J. Pople. 1991. Theoretical study of blocked glycine and alanine peptide analogs. *J. Am. Chem. Soc.* 113:5989–5997.

294. Vargas, R., J. Garza, B. Hay, and D. Dixon. 2002. Conformational study of the alanine dipeptide at the MP2 and DFT levels. *J. Phys. Chem. A* 106:3213–3218.

295. Ramachandran, G.N., and V. Sasisekharan. 1968. Conformation of polypeptides and proteins. *Adv. Protein Chem.* 23: 283–438.

296. Zimmerman, S.S., M. Pottle, G. Nemethy, and H. Scheraga. 1977. Conformational analysis of the 20 naturally occurring amino acid residues using ECEPP. *Macromolecules* 10:1–9.

297. In, Y., M. Fujii, Y. Sasada, and T. Ishida. 2001. Structural studies on C-amidated amino acids and peptides: Structures of hydrochloride salts of C-amidated Ile, Val, Thr, Ser, Met, Trp, Gln and Arg, and comparison with their C-unamidated counterparts. *Acta. Crystallogr. B* 57:72–81.

298. Burns, C.S., E. Aronoff-Spencer, C.M. Dunham, P. Lario, N.I. Avdievich, W.E. Antholine, M.M. Olmstead, A. Vrielink, G.J. Gerfen, J. Peisach, W.G. Scott, and G.L. Millhauser. 2002. Molecular features of the copper binding sites in the octarepeat domain of the prion protein. *Biochemistry* 41:3991–4001.

299. Karle, I.L. 1980. Erythro-β-hydroxy-N-acetyl-tryptophanamide benzene solvate. *ACA*, Ser. 22, 7:25–27.

300. Souhassou, M., A. Aubry, and C. Lecomte. 1990. Structure of N-acetyl-N-methyl-L-tryptophanamide. *Acta Crystallogr. C* 46:1303–1305.

301. Ichikawa, T., and Y. Iitaka. 1969. The crystal structure of DL-acetylleucine N-methylamide, $C_9H_{18}O_2N_2$. *Acta Crystallogr. B* 25:1824–1833.

302. Puliti, R., C.A. Mattia, C. Giancola, and G. Barone. 2000. Crystal structure and conformational stability of N-acetyl-prolyl-leucinamide. Comparison between structural and thermophysical data. *J. Mol. Struct.* 553:117–130.

303. Jouvet, M. 1999. Sleep and serotonin: An unfinished story. *Neuropsychopharmacol.* 21:S24–S27.

304. Philips, L.A., and D.H. Levy. 1988. Rotationally resolved electronic spectroscopy of tryptamine conformers in a supersonic jet. *J. Chem. Phys.* 89:85–89.

305. Philips, L.A., S. Webb, S. Martinez, G. Fleming, and D. Levy. 1988. Time-resolved spectroscopy of tryptophan conformers in a supersonic jet. *J. Am. Chem. Soc.* 110:1352–1355.

306. Peteanu, L.A., and Levy, D.H. 1988. Spectroscopy of complexes of tryptamine and 3-indolepropionic acid with various solvents. *J. Phys. Chem.* 92:6554–6561.
307. Connell, L.L., T. Corcoran, P. Joireman, and P. Felker. 1990. Conformational analysis of jet-cooled tryptophan analogs by rotational coherence spectroscopy. *Chem. Phys. Lett.* 166:510–516.
308. Pauwels, P.J., and G.W. John. 1999. Present and future of 5-HT receptor agonists as antimigraine drugs. *Clin. Neuropharm.* 22:123–136.
309. Carney, J.R., and T. Zwier. 2000. The infrared and ultraviolet spectra of individual conformational isomers of biomolecules: Tryptamine. *J. Phys. Chem.* 104A:8677–8688.
310. Carney, J.R., and T. Zwier. 2001. Conformational flexibility in small biomolecules: Tryptamine and 3-indole-propionic acid. *Chem. Phys. Lett.* 341:77–85.
311. Caminati, W. 2004. The rotational spectra of conformers of biomolecules: Tryptamine. *Phys. Chem. Chem. Phys.* 6:2806–2810.
312. Schmitt, M., M. Bohm, C. Ratzer, C. Vu, L. Kalkman, and W. Meerts. 2005. Structural selection by microsolvation: conformational locking of tryptamine. *J. Am. Chem. Soc.* 127:10356–10364.
313. Schmitt, M., C. Marian, S. Salzmann, and W. Meerts. 2006. Electronically excited states of tryptamine and its microhydrated complex. *J. Chem. Phys.* 125:124309.
314. Schmitt, M., M. Boehm, and K. Kleinermanns. 2006. Low-frequency backbone vibrations of individual conformational isomers: Tryptamine. *J. Chem. Phys.* 125:144303.
315. Nguyen, T.V. 2006. Permanent electric dipole moments of four tryptamine conformers in the gas phase: A new diagnostic of structure and dynamics. *J. Chem. Phys.* 124:054317.
316. Nguyen, T.V., and D.W. Pratt. 2005. Tryptamine in the gas phase. A high resolution laser study of the structural and dynamic properties of its ground and electronically excited states. *Mol. Phys.* 103:1603–1613.
317. Clarkson, J., B. Dian, L. Moriggi, V. McCarthy, A. DeFusco, K. Jordan, and T.S. Zwier. 2008. Direct measurement of the energy thresholds to conformational isomerization in tryptamine: Experiment and theory. *J. Chem. Phys.* 122:214311.
318. Zhu, Z., A. Munhall, and St. Johnson. 2007. Tyramine excites rat subthalamic neurons *in vitro* by a dopamine-dependent mechanism. *Neuropharmacol.* 52:1169–1178.
319. Ishida, T., and M. Inoue. 1981. Structure of the complex 3-indoleacetic acid: Tyramine (1:1). *Acta Crystallogr.* 37B:2117–2120.
320. Tamura, K., A. Wakahara, T. Fujiwara, and K.-I. Tomita. 1974. The crystal and molecular structure of tyramine hydrochloride. *Bull. Chem. Soc. Japan* 47:2682–2685.
321. Ogawa, K., K. Tago, T. Ishida, and K.-I. Tomita. 1980. The structure of the 1-thyminylacetic acid and tyramine (1:1) complex. *Acta Crystallogr.* 36B:2095–2099.
322. Zakaria, C.M., G. Ferguson, A.J. Lough, and C. Glidewell. 2002. Ferrocene-1,1′-dicarboxylic acid as a building block in supramolecular chemistry: Supramolecular structures in one, two and three dimensions. *Acta Crystallogr.* 58B:786–802.
323. Yoon, I., K. Seo, S. Lee, Y. Lee, and B. Kim. 2007. Conformational study of tyramine and its water clusters by laser spectroscopy. *J. Phys.Chem.* 111A:1800–1807.
324. Pillsbury, N., and T.S. Zwier. 2009. Conformation-specific spectroscopy and excited state photophysics of 5-phenyl-1-pentene. *J. Phys. Chem.* 113A:118–125.
325. Choi, Y., and D. Lubman. 1992. Analytical spectroscopy and structure of biomolecules using an *ab initio* computational method. *Anal. Chem.* 64:2726–2734.
326. Zwier, T. 2001. Laser spectroscopy of jet-cooled biomolecules and their water-containing clusters: Water bridges and molecular conformation. *J. Phys. Chem.* 105A:8827–8839.
327. Zwier, T. 2006. Laser probes of conformational isomerization in flexible molecules and complexes. *J. Phys. Chem.* 110A:4133–4150.

328. LeGreve, T., A. Baquero, and T.S. Zwier. 2007. Infrared and ultraviolet spectral signatures and conformational preferences of jet-cooled serotonin. *J. Am. Chem. Soc.* 129:4028–4038.

329. Melandri, S., and A. Maris. 2004. Intramolecular hydrogen bonds and conformational properties of biogenic amines: A free-jet microwave study of tyrami. *Phys. Chem. Chem. Phys.* 6: 2863–2866.

330. Kolev, T., M. Spiteller, T. van Almsick, H. Mayer-Figger, and W.S. Sheldrick. 2007. 2-(Indol-3-yl)ethylammonium dihydrogenphosphate. *Acta Crystallogr.* E63:o179–o181.

331. Nowell, H., J.P. Attfield, and J.C. Cole. 2002. The use of restraints in Rietveld refinement of molecular compounds: A case study using the crystal structure determination of tryptamine free base. *Acta Crystallogr.* 58B:835–840.

332. Terakita, A., H. Matsunaga, T. Ueda, T. Eguchi, M. Echigoya, K. Umemoto, and M. Godo. 2004. Investigation of intermolecular interaction in molecular complex of tryptamine and benzoic acid by solid-state 2D NMR. *Chem. Pharmaceut. Bull.* 52:546–551.

333. Kamiichi, K., M. Danshita, N. Minamino, M. Doi, T. Ishida, and M. Inoue. 1986. Indole ring binds to 7-methylguanine base by $\pi$-$\pi$ stacking interaction: Crystal structure of 7-methylguanosine 5′-monophosphate-tryptamine complex. *FEBS Letters.* 195:57–60.

334. Koshima, H., M. Nagano, and T. Asahi. 2005. Optical activity induced by helical arrangements of tryptamine and 4-chlorobenzoic acid in their cocrystal. *J. Am. Chem. Soc.* 127:2455–2463.

335. Siripaisarnpipat, S., and S. Larsen. 1987. The crystal structure at 105 K of two compounds derived from tryptamine, 1-trifluoroacetyl-3-(2-trifluoroacetamidoethyl)indole and tryptamine 3,4-dimethoxybenzoate. *Acta Chem. Scand.* 41:539–547.

336. Koshima, H., S.I. Khan, and M.A. Garcia-Garibay. 1998. Chiral crystalline salts from achiral biphenylcarboxylic acids and tryptamine. *Tetrahedron Asym.* 9:1851–1854.

337. Koshima, H., and M. Miyauchi. 2001. Polymorphs of a cocrystal with achiral and chiral structures prepared by pseudoseeding: tryptamine/hydrocinnamic acid. *Cryst. Growth Des.* 1:355–357.

338. Bartland, G., G. Freeman, and C. Bugg. 1974. Crystal structures of tryptamine picrate and D,L-tryptophan picrate-methanol, two indole donor-acceptor complexes. *Acta Crystallogr.* 30B:1841–1849.

339. Ito, Y., and H. Fujita. 2000. Unusual [2+2] Photocycloaddition between tryptamine and 3-nitrocinnamic acid in the solid state. *Chem. Lett.* 29:288–289.

340. Inoue, M., T. Sakaki, T. Fujiwara, and K. Tomita. 1978. Structural studies of the interaction between indole derivatives and biologically important aromatic compounds. II. The crystal and molecular structure of the tryptamine–phenylacetic acid (1:1) complex. *Bull. Chem. Soc. Japan* 51:1123–1127.

341. Wakahara, A., T. Fujiwara, and K.-I. Tomita. 1973. Structural studies of tryptophan metabolites by X-ray diffraction method. II. The crystal and molecular structure of tryptamine hydrochloride. *Bull. Chem. Soc. Japan* 46:2481–2486.

342. Koshima, H., S. Honke, and J. Fujita. 1999. Generation of chirality in two-component molecular crystals of tryptamine and achiral carboxylic acids. *J. Org. Chem.* 64:3916–3921.

343. Martino, P., P. Conflant, M. Drache, J. Huvenne, and A. Hermann. 1997. Preparation and physical characterization of forms II and III of paracetamol. *J. Thermal Anal. Calor.* 48:447–458.

344. Szelagiewicz, M., C. Marcolli, S. Cianferani, A. P. Hard, A. Vit, A. Burkhard, M. von Raumer, U.Ch. Hofmeier, A. Zilian, E. Francotte, and R. Schenker. 1999. *In situ* characterization of polymorphic forms: the potential of Raman techniques. *J. Thermal Anal. Calor.* 57:23–43.

345. Haisa, M., S. Kashino, R. Kawai, and H. Maeda. 1976. The monoclinic form of p-hydroxyacetanilide. *Acta Crystallogr. B* 32:1283–1285.

346. Haisa, M., S. Kashino, and H. Maeda. 1974. The orthorhombic form of p-hydroxyacetanilide. *Acta Crystallogr. B* 30:2510–2512.

347. Naumov, D., M.A. Vasilchenko, and K.A. Howard. 1998. The monoclinic form of acetaminophen at 150 K. *Acta Crystallogr. C* 54:653–655.

348. Beyer, T., G.M. Day, and S.L. Price. 2001. The prediction, morphology, and mechanical properties of the polymorphs of paracetamol. *J. Am. Chem. Soc.* 123:5086–5094.

349. Boldyreva, E.V., T. Shakhtshneider, H. Ahsbahs, H. Sowa, and H. Uchtmann. 2002. DSC and adiabatic calorimetry study of the polymorphs of paracetamol. *J. Thermal Anal. Calor.* 68:437–452.

350. Tantishaiyakul, V., N. Phadoongsombut, S. Kamaung, S. Wongwisansri, and P. Mathurod. 1999. Fourier transform infrared spectrometric determination of paracetamol and ibuprofen in tablets. *Pharmazie* 54:111–117.

351. Binev, I.G., P. Vasssileva-Boyadjieva, and Y.I. Binev. 1998. Experimental and *ab initio* MO studies on the IR spectra and structure of 4-hydroxyacetanilide (paracetamol), its oxyanion and dianion. *J. Mol. Struct.* 447:235–246.

352. Freedman, H. H. 1960. Electronic interactions in phenyl acetates and acetanilides. *J. Am. Chem. Soc.* 82:2454–2459.

353. Ran, A. 1993. Basic experiments in thin-layer chromatography–Fourier transform Raman spectrometry. *J. Raman Spectrosc.* 24:251–254.

354. Dressi, E., G. Germelli, P. Corti, M. Massacesi, and P. Perrucio. 1995. Quantitative Fourier transform near-infrared spectroscopy in the quality control of solid pharmaceutical formulations. *Analyst* 120:2361–2367.

355. Boldyreva, E.V. 2003. High-pressure studies of the anisotropy of structural distortion of molecular crystals. *J. Mol. Struct.* 647:159–179.

356. Tawashi, R. 1968. Aspirin: dissolution rates of two polymorphic forms. *Science* 160:76–80.

357. Summers, M.P., R.P. Enever, and J.E. Carless. 1973. The polymorphism of aspirin, in: *Particle Growth in Suspensions*, A. Smith, (Ed.). Academic Press, London.

358. Rowe, R.C., R.J. Roberts, M.H. Charlton, and R. Docherty. 1999. Potential polymorphs of aspirin. *J. Comput. Chem.* 20:262–273.

359. Vishweshwar, P., J.A. McMahon, M. Oliveira, M.L. Peterson, and M.J. Zaworotko. 2005. The predictably elusive form II of aspirin. *J. Am. Chem. Soc.* 127:16802–16803.

360. Bond, A., R. Böse, and G.R. Desiraju. 2007. Zur Polymorphie von Aspirin: kristallines Aspirin als zwei ineinander verwachsene polymorphe Domäne. *Angew. Chem.* 119:625–630.

361. Bond, A., R. Böse, and G.R. Desiraju. 2007. Zur Polymorphie von Aspirin. *Angew. Chem.* 119:621–624.

362. Thompson, H.W., D.L. Nicholson, and L.N. Short. 1950. Vibrational spectra of complex molecules. The infra-red spectra of complex molecules. *Discuss. Faraday Society* 9:222–235.

363. Dziegielewski, J., J. Hanuza, B. Ezowska-Tsebiatowska, and I.Z. Siemion. 1973. Infrared spectra and structure of some penicillin derivatives. *J. Appl. Spectrosc.* 19:1031–1038.

364. Kupka, T. 1997. β-Lactam antibiotics. Spectroscopy and molecular orbital (MO) calculations: Part I: IR studies of complexation in penicillin-transition metal ion systems and semi-empirical PM3 calculations on simple model compounds. *Spectrochim. Acta. A* 53:2649–2658.

365. Nangia, A., and G.R. Desiraju. 1999. Axial and equatorial conformations of penicillins, their sulphoxides and sulphones: the role of N···HS and C···HO hydrogen bonds. *J. Mol. Struct.* 474:65–79.

366. Ramaswamy, S., R.K. Rajaram, and V. Ramakrishnan. 2002. Raman and IR spectral studies of D-phenylglycinium perchlorate. *J. Raman Spectrosc.* 33:689–698.

367. García-Reiriz, A., P.C. Damiani, and. A.C. Olivieri. 2007. Analysis of amoxicillin in human urine by photo-activated generation of fluorescence excitation–emission matrices and artificial neural networks combined with residual bilinearization. *Anal. Chim. Acta* 588:192–199.

368. Chulavatnatol, S., and B. Charles. 1993. High-performance liquid chromatographic determination of amoxicillin in urine using solid-phase, ion-pair extraction and ultraviolet detection. *J. Chromatogr.* 615:91–96.

369. Iliescu, T., M. Baia, and I. Pavel. 2006. Raman and SERS investigations of potassium benzylpenicillin. *J. Raman Spectrosc.* 37:318–325.

370. Ghassempour, A., S. Davarani, M. Noroozi, and M. Shamsipur. 2005. Determination of ternary mixtures of penicillin G, benzathine and procaine by liquid chromatography and factorial design study. *Talanta* 65:1038–1044.

371. Murillo, J.A., J. Rodríguez, J.M. Lemus, and A. Alañón. 1990. Determination of amoxicillin and cephalexin in mixtures by second-derivative spectrophotometry. *Analyst* 115:1117–1119.

372. Krauwinkel, W., N. Volkers-Kamermans, and J. van Zijtveld. 1993. Determination of amoxicillin in human plasma by high-performance liquid chromatography and solid phase extraction. *J. Chromatogr.* 617:334–338.

373. Gil García, M.D., M.J. Culzoni, M.M. De Zan, R. Santiago Valverde, M. Martínez Galera, and H.C. Goicoechea. 2008. Solving matrix effects exploiting the second-order advantage in the resolution and determination of eight tetracycline antibiotics in effluent wastewater by modelling liquid chromatography data with multivariate curve resolution-alternating least squares and unfolded-partial least squares followed by residual bilinearization algorithms: II. Prediction and figures of merit. *J. Chromatogr. A* 1179:115–124.

374. Morelli, B. 1987. Spectrophotometric assay for chloramphenicol and some derivatives in the pure form and in formulations. *J. Pharmaceut. Biomed. Anal.* 5:577–583.

375. Montaudo, G., S. Caccamese, V. Librando, and A. Recca. 1975. Structural analysis by lanthanide shift reagents: IV. Conformation of the amide group as a function of the ring size in lactams. *J. Mol. Struct.* 27:303–308.

376. Hallam, H.E., and C.E. Jones. 1970. Conformational isomerism of the amide group—A review of the IR and NMR spectroscopic evidence. *J. Mol. Struct.* 5:1–19.

377. Raber, D., Ch.M. Janks, M.D. Johnston Jr., and N.K. Raber. 1980. Structure elucidation with lanthanide induced shifts. 9. bicyclo[3.3.1]nonan-9-one. *Tetrahedron Lett.* 27: 677–680.

378. Al-Zoubi, N., J.E. Koundourellis, and S. Malamataris. 2002. FT-IR and Raman spectroscopic methods for identification and quantitation of orthorhombic and monoclinic paracetamol in powder mixes. *J. Pharmaceut. Biomed. Anal.* 29:459–467.

379. Selling, H.A. 1971. Assessment of configuration of four isomeric α,β-unsaturated sulfones using a lanthanide shift reagent. *Tetrahedron* 31:2543–2546.

380. Kim, Y., K. Machida, T. Taga, and K. Osaki. 1985. Structure redetermination and packing analysis of aspirin crystal. *Chem. Pharm. Bull.* 33:2641–2647.

381. Koleva, B., M. Spiteller, and T. Kolev. 2010. Polarized spectroscopic elucidation of N-acetyl-l-cysteine, l-cysteine, l-cystine, l-ascorbic acid and a tool for their determination in solid mixtures. *Amino Acids* 38:295–304.

382. Krishna, H., M. Vijayan, and L. Brehm. 1979. Sodium((2,3-dihydro-1,5-dimethyl-3-oxo-2-phenyl-1H-pyrazol-4-yl)methylamino)-methanesulfonate monohydrate. *Acta Crystallogr. B* 35:612–613.

383. Seidel, R.W., R. Bakalska, T. Kolev, D. Vassilev, H. Mayer-Figge, M. Spiteller, W.S. Sheldrick, and B.B. Koleva. 2009. N-methylcodeinium iodide—Crystal structure and spectroscopic elucidation. *Spectrochim. Acta* 73:61–66.

384. Marder, S., J. Perry, B.G. Tiemann, R.E. Marsh, and W.P. Schaefer. 1990. Second-order optical nonlinearities and photostabilities of 2-N-methylstilbazolium salts. *Chem. Mater.* 2:685–690.

385. Ashwell, G., R. Hargreaves, C.E. Baldwin, G. Bahra, and C.R. Brown. 1992. Improved second-harmonic generation from Langmuir–Blodgett films of hemicyanine dyes. *Nature* 357:393–395.

386. Okada, S., T. Taniuchi, and H. Nakanishi. 2007. Preparation of organic crystals for coherent terahertz-wave generation. *Mol. Electron Bioelectron* 18:89–92.

387. Okada, S. 2006. Organic nonlinear optical materials, in: *Handbook of Electronic Materials*, T. Kimura, T. Yao, T. Okumura, and T. Toyota (Eds.). Asakura, Tokyo.

388. Taniuchi, T., S. Okada, and H. Nakanishi. 2006. Development of an organic crystalline material toward terahertz-wave applications. *Sci. Mac.* 58:823–829.

389. Okada, S., T. Taniuchi, and H. Nakanishi. 2006. Generation of unexplored light from organic crystals: coherent and broad-band terahertz-wave generation by difference frequency generation. *Opt. Alliance* 17:14–17.

390. Okada, S. 2000. Nonlinear optical crystals, in: *Recent Advances in Research and Development of Organic Crystalline Materials*, H. Nakanishi (Ed.). CMC, Tokyo.

391. Kanno, S., and S. Okada. 2007. Synthesis of phenylbutadiynylpyridinium derivatives for nonlinear optics. *Mol. Cryst. Liq. Cryst.* 471:365–371.

392. Mineno, Y., T. Matsukawa, S. Okada, S. Ikeda, T. Taniuchi, H. Nakanishi, H. Adachi, M. Yoshimura, Y. Mori, and T. Sasaki. 2007. Single crystal preparation of DAST for terahertz-wave generation. *Mol. Cryst. Liq. Cryst.* 463:55–61.

393. Matsukawa, T., Y. Mineno, T. Odani, S. Okada, T. Taniuchi, and H. Nakanishi. 2007. Synthesis and terahertz-wave generation of mixed crystals composed of 1-methyl-4-{2-[4-(dimethylamino)phenyl]ethenyl}pyridinium p-toluenesulfonate and p-chlorobenzenesulfonate. *J. Cryst. Growth* 299:344–348.

394. Kwon, E., S. Okada, and H. Nakanishi. 2007. Relationship between THz energy decay and molecular vibration of 1-methyl-4-{2-[4-(dimethylamino)phenyl]ethenyl}pyridinium p-toluenesulfonate derivatives. *Japan J. Appl. Phys.* 46:146–148.

395. Taniuchi, T., S. Ikeda, Y. Mineno, S. Okada, and H. Nakanishi. 2005. Terahertz properties of a new organic crystal, 4′-dimethylamino-n-methyl-4-stilbazolium-chlorobenzenesulfonate. *Japan J. Appl. Phys.* 44:L932–L934.

396. Glavcheva, Z., H. Umezawa, Y. Mineno, T. Odani, S. Okada, S. Ikeda, T. Taniuchi, and H. Nakanishi. 2005. Synthesis and properties of 1-methyl-4-{2-[4-(dimethylamino)phenyl]ethenyl}pyridinium p-toluenesulfonate derivatives with isomorphous crystal structure. *Japan J. Appl. Phys.* 44:5231–5235.

397. Marder, S. R., J. W. Perry, and C. P. Yakymyshyn. 1994. Organic salts with large second-order optical nonlinearities. *Chem. Mater.* 6:1137–1147.

398. Yang, Z., M. Wörle, L. Mutter, M. Jazbinsek, and P. Günter. 2007. Synthesis, crystal structure, and second-order nonlinear optical properties of new stilbazolium salts. *Cryst. Growth Des.* 7:83–86.

399. Yang, Z., L. Mutter, M. Stillhart, B. Ruiz, S. Aravazhi, M. Jazbinsek, A. Schneider, V. Gramlich, and P. Günter. 2007. Large-size bulk and thin-film stilbazolium-salt single crystals for nonlinear optics and THz generation. *Adv. Fun. Mater.* 17:2018–2023.

400. Yang, Z., M. Jazbinsek, B. Ruiz, S. Aravazhi, V. Gramlich, and P. Günter. 2007. Molecular engineering of stilbazolium derivatives for second-order nonlinear optics. *Chem. Mater.* 19:3512–3518.

401. Ruiz, B., Z. Yang, V. Gramlich, M. Jazbinsek, and P. Günter. 2006. Synthesis and crystal structure of a new stilbazolium salt with large second-order optical nonlinearity. *J. Mater. Chem.* 16:2839–2842.

402. Yang, Z., S. Aravazhi, A. Schneider, P. Seiler, M. Jazbinsek, and P. Günter. 2005. Synthesis and crystal growth of stilbazolium derivatives for second-order nonlinear optics. *Adv. Fun. Mater.* 15:1072–1076.

403. Mutter, L., M. Jazbinsek, M. Zgonik, U. Meier, Ch. Bosshard, and P. Günter, 2003. Photobleaching and optical properties of organic crystal 4-n, n-dimethylamino-4'-n'-methyl stilbazolium tosylate. *J. Appl. Phys.* 94:1356–1361.

404. Follonier, S., M. Fierz, I. Biaggio, U. Meier, Ch. Bosshard, and P. Günter. 2002. Structural, optical, and electrical properties of the organic molecular crystal 4-n, n-dimethylamino-4'-n'-methyl stilbazolium tosylate. *J. Opt. Soc. Am. B* 19:1990–1998.

405. Serbutoviez, C., J.F. Nicoud, J. Fischer, I. Ledoux, and Z. Zyss. 1994. Crystalline zwitteronic stilbazolium derivatives with large quadratic optical nonlinearities. *Chem. Mater.* 8:1360–1367.

406. Schneider, A., M. Neis, M. Stillhart, B. Ruiz, R. Khan, and P. Günter. 2006. Generation of terahertz pulses through optical rectification in organic DAST crystals: Theory and experiment. *J. Opt. Soc. Am. B* 23:1822–1835.

407. Mutter, L., P. Dittrich, M. Jazbinsek, and P. Günter. 2004. Growth and planar structures of DAST crystals for optical applications. *J. Non. Opt. Phys. Mater.* 13:559–567.

408. Lavéant, P., C. Medrano, B. Ruiz, and P. Günter. 2003. Growth of nonlinear optical DAST crystals. *Chimia* 57:349–351.

409. Schneider, A., I. Biaggio, and P. Günter. 2003. Optimized generation of THz pulses via optical rectification in the organic salt DAST. *Opt. Comm.* 224:337–341.

410. Bosshard, Ch., R. Spreiter, L. Degiorgi, and P. Günter. 2002. Infrared and Raman spectroscopy of the organic crystal DAST: Polarization dependence and contribution of molecular vibrations to the linear electro-optic effect. *Phys. Rev. B* 66:205107.1–205107.9.

411. Meier, U., M. Bösch, Ch. Bosshard, and P. Günter. 1999. Phase matched parametric interactions in DAST at telecommunication wavelengths. *Nonlinear Opt.* 22:279–282.

412. Spreiter, R., Ch. Bosshard, F. Pan, and P. Günter. 1997. High-frequency response and acoustic phonon contribution of the linear electro-optic effect in DAST. *Opt. Lett.* 22:564–566.

413. Pan, F., M.S. Wong, Ch. Bosshard, and P. Günter. 1996. Crystal growth and characterization of the organic salt 2-N,N-dimethylamino-4'-N'-methyl-stilbazolium tosylate (DAST). *Adv. Mater.* 8:592–595.

414. Coe, B. 2006. Switchable nonlinear optical metallochromophores with pyridinium electron acceptor groups. *Acc. Chem. Res.* 39:383–393.

415. Yang, M., and B. Champagne. 2003. Large off-diagonal contribution to the second-order optical nonlinearities of λ-shaped molecules. *J. Phys. Chem. A* 107:3942–3951.

416. Kolev, T.M., D.Y. Yancheva, and S.I. Stoyanov. 2004. Synthesis, spectral and structural elucidation of some pyridinium betaines of squaric acid—Potential materials for nonlinear optical applications. *Adv. Func. Mater.* 14:799–804.

417. Kolev, T., R. Wortmann, M. Spiteller, W.S. Sheldrick, and M. Heller. 2004. 4-Phenylpyridinium hydrogensquarate. *Acta Crystallogr. E* 60:o956–o958.

418. Kolev, T., R. Wortmann, M. Spiteller, W.S. Sheldrick, and M. Heller. 2004. 4-Methoxypyridinium betaine of squaric acid. *Acta Crystallogr. E* 60:o1376–o1376.

419. Kolev, T., I. Kityk, J. Ebothe, and B. Sahraoui. 2007. Intrinsic hyperpolarizability of 3-dicyanomethylene-5,5-dimethyl-1-[2-(4-hydroxyphenyl)ethenyl)]-cyclohexene nanocrystallites incorporated into the photopolymer matrices. *Chem. Phys. Lett.* 443:309–312.

420. Koleva, T., B. Koleva, J. Kasperczyk, I. Kityk, S. Tkaczyk, M. Spiteller, A.H. Reshak, and W. Kuznik. 2009. Novel nonlinear optical materials based on dihydropyridine organic chromophore deposited on mica substrate. *J. Mater. Sci.: Mater. Electron.* 20:1073–1077.

421. Kolev, T., D. Kleb, D.Yancheva, M. Schürmann, H. Preut, and P. Bleckmann. 2001. Crystal structure of cesium 4,6-dinitroresorcinolate, $CsC_6H_3O_2(NO_3)_{(2)}$. *Zeitschrift für Kristallographie* 216:63–64.

422. Kolev, T., Z. Glavcheva, D. Yancheva, M. Schurmann, D. Kleb, H. Preut, and P. Bleckmann. 2001. 2-{3-[2-(4-Hydroxyphenyl)vinyl]-5,5-dimethylcyclo-hex-2-en-1-ylidene}malononitrile. *Acta Crystallogr. E* 57:o561–o562.

423. Kolev, T., Z. Glavcheva, M. Schurmann, D. Kleb, H. Preut, and P. Bleckmann. 2001. 2-{5,5-Dimethyl-3-[2-(2-methoxyphenyl)vinyl]cyclo-hex-2-enylidene}malononitrile. *Acta Crystallogr. E* 57:o1166–o1167.

424. Kolev, T., Z. Glavcheva, D. Yancheva, M. Schurmann, D. Kleb, H. Preut, and P. Bleckmann. 2001. 2-{3-[2-(1H-Indol-3-yl)vinyl]-5,5-dimethylcyclohex-2-enylidene}malononitrile. *Acta Crystallogr. E* 57:o760–o761.

425. Kolev, T., Z. Glavcheva, D. Yancheva, M. Schurmann, D. Kleb, H. Preut, and P. Bleckmann. 2001. Triclinic form of 2-{5,5-dimethyl-3-[2-(2,4,6-trimethoxyphenyl)vinyl]cyclohex-2-enylidene}malononitrile. *Acta Crystallogr. E* 57:o966–o967.

426. Kolev, T., Z. Glavcheva, M. Schurmann, M. Kleb, H. Preut, and P. Bleckmann. 2001. Monoclinic form of 2-{5,5-dimethyl-3-[2-(2,4,6-trimethoxyphenyl)vinyl]cyclohex-2-enylidene}malononitrile. *Acta Crystallogr. E* 57:o964–o965.

427. Kolev, T., D. Yancheva, B. Shivachev, R. Petrova, and M. Spiteller. 2005. 2-[3-[(3,4-Dimethoxyphenyl)ethenyl]-5,5-dimethylcyclohex-2-enylidene]-malononitrile. *Acta Crystallogr. E* 61:o550–o552.

428. Kolev, T., M. Schurmann, D. Kleb, H. Preut, and M. Spiteller. 2002. 2-{3-[2-(2,4-Dimethoxyphenyl)vinyl]-5,5-dimethyl-cyclohex-2-enylidene}malononitrile. *Acta Crystallogr. E* 58:o1172–o1173.

429. Ashwell, G.J. 1999. Langmuir-Blodgett films: Molecular engineering of non-centrosymmetric structures for second-order nonlinear optical applications *J. Mater. Chem.* 9:1991–1996.

430. Metzger, R.M., R. Chen, U. Hopfner, M. Lakshmikantham, D. Vuillaume, T. Kawai, X. Wu, H. Tachibana, T. Hughes, H. Sakurai, J. Baldwin, C. Hosch, B. Cava, L. Brehmer, and G. Ashwell. 1997. Unimolecular electrical rectification in hexadecylquinolinium tricyanoquinodimethanide. *J. Am. Chem. Soc.* 119:10455–10466.

431. Wolff, J.J., and R. Wortmann. 1999. Organic materials for second-order non-linear optics, in: *Advances in Physical Organic Chemistry*, Vol. 32, D. Bethell (Ed.). Academic Press, London.

432. Würthner, F., and S. Yao. 2000. Dipolar dye aggregates: A problem for nonlinear optics, but a chance for supramolecular chemistry. *Angew. Chem. Int. Ed.* 39:1978–1980.

433. Yao, S, U. Beginn, T., Gress, M. Lysetska, and F. Würthner. 2004. Supramolecular polymerization and gel formation of bis(merocyanine) dyes driven by dipolar aggregation. *J. Am. Chem. Soc.* 126:8336–8348.

434. Kolev, T., T. Tsanev, S. Kotov, H. Mayer-Figge, M. Spiteller, W.S. Sheldrick, and B. Koleva. 2009. Anyles of 4-benzoylpyridine—Crystal structure and spectroscopic properties. *Dyes Pigm.* 82:95–101.

435. Koleva, B.B., T. Kolev, M. Spiteller, H. Mayer-Figge, and W.S. Sheldrick. 2009. 1-methyl-4-[2-(3-methoxy-4-hydroxyphenyl)ethenyl)]pyridinium] hydrogensquarate monohydrate—Crystal structure and spectroscopic properties. *J. Inclus. Phenom.* 64:173–181.

436. Würthner, F., S. Yao, B. Heise, and C. Tschierskec. 2001. Hydrogen bond directed formation of liquid-crystalline merocyanine dye assemblies. *Chem. Comm.* 2001:2260–2261.

437. Würthner, F., S. Yao, and U. Beginn. 2003. Highly ordered merocyanine dye assemblies by supramolecular polymerization and hierarchical self-organization. *Angew. Chem. Int. Ed.* 42:3247–3249.

438. Gruselle, M., B. Malezieux, S. Benard, C. Train, C. Guyard-Duhayon, P. Gredin, K. Tonsuaadu, and R. Clement. 2004. Chiral matrix effect of optically active oxalate-based networks: Controlled helical conformation of an organic chromophore. *Tetrahedron Asymm.* 15:3103–3109.

439. Cariati, E., R. Ugo, F. Cariati, D. Roberto, N. Masciocchi, S. Galli, and A. Sironi. 2001. J-aggregates granting giant second-order NLO responses in self-assembled hybrid inorganic-organic materials. *Adv. Mater.* 13:1665–1668.

440. Tan, X., S. Sun, W. Yu, D. Xing, Y. Wang, and C. Qi. 2004. trans-4-[p-(N,N-Diethylamino)styryl]-N-methylpyridinium triiodide. *Acta Crystallogr. E* 60:o1054–o1056.

441. Benard, S., P. Yu, J.P. Audiere, E. Riviere, R. Clement, J. Guilhem, L. Tchertanov, and K. Nakatani. 2000. Structure and NLO properties of layered bimetallic oxalato-bridged ferromagnetic networks containing stilbazolium-shaped chromophores. *J. Am. Chem. Soc.* 122:9444–9454.

442. Ren, Y., Q. Fang, W. Yu, H. Lei, Y. Tian, M. Jiang, Q. Yang, and T.C.W. Mak. 2000. Synthesis, structures and two-photon pumped up-conversion lasing properties of two new organic salts. *J. Mater. Chem.* 10:2025–2031.

443. Varsanyi, G. 1969. *Vibrational Spectra of Benzene Derivatives.* Academiai Kiado, Budapest.

444. Lemke, R. 1970. Solvatochromie von 80 µm in verschiedenen Alkoholen bei Arylidenisophoron-Abkömmlingen. *Chem. Ber.* 103:1894–1899.

445. Kolev, T., D. Yancheva, and B. Stamboliyska. 2003. Experimental and computational studies of the structure and vibrational spectra of 2-[5,5-dimethyl-3-(2-phenyl-vinyl)cyclohex-2-enylidene]-malononitrile. *Spectrochim. Acta A* 59:3325–3335.

446. Pariente, G.L., and P.R. Griffiths. 1986. Introduction to spectral deconvolution. *Trends Anal. Chem.* 5:209–215.

447. Griffiths, P.R., and G.L. Pariente. 1988. Fourier transform infrared spectrometry—A tool for modern agricultural research, in: *Instrumentation for the 21st Century*, G. Beecher (Ed.). Martins Nijhoff, Boston, pp. 223–246.

# Index

Result spectrum, 46
Rigid nucleus, 15
Rotation axis of symmetry, 9

**S**

Salophen, 124, 136
Sample spectrum, 46
Saturn-ring defects, 36
Savitzky–Golay (method), 31, 46, 47, 48
Second-harmonic generation (SHG), 99, 151
Serotonin, 113, 114
SHG; *See* Second-harmonic generation
Small peptides, 73, 110, 149
Smoothing, 31, 45, 46, 49
Spectral subtract, 21
Squaric acid, 38, 39, 40, 41, 103, 110, 113, 148
Stepwise reduction method, 21
Stilbazolium salts, 73, 151, 166
Subtracted spectrum, 46
Subtraction, 2, 18, 21, 46, 53
Subtraction factor, 46
Surface tension coefficient, 33
Symmetry class, 11, 12, 22, 55, 175

**T**

TD-DFT; *See* Time-dependence density
        functional theory
Thermogravimetric analysis (TGA), 99
Thermotropic, 15

Thiobarbituric derivatives, 145, 149
Time-dependence density functional theory
        (TD-DFT), 87–88
Tryptamine, 114, 120
Tryptammonium (salts), 118, 119, 120, 122
Tryptophan, 107, 112, 114
L-Tryptophanamide, 105, 106, 110, 111

**U**

UHF theory; *See* Unrestricted Hartree–Fock
        theory
Ultraviolet-visible spectroscopy
        (UV-VIS), 151
Unrestricted Hartree–Fock (UHF) theory, 81
UV-VIS; *See* Ultraviolet-visible
        spectroscopy

**V**

Validation, 31, 36, 45, 46, 136
Vibrational analysis, 172
Violurate (salts), 73

**X**

X-ray diffraction (XRD), 73, 123, 151

**Z**

Zero-crossing technique of measurement, 126

Milton Keynes UK
Ingram Content Group UK Ltd.
UKHW022039141024
449569UK00014B/663